Handbooks for the Identification of British Insects
Vol. 1 Part 17

# The adult Trichoptera (caddisflies) of Britain and Ireland

Peter Barnard and Emma Ross
c/o Department of Entomology
The Natural History Museum
Cromwell Road
London SW7 5BD

Published for the Royal Entomological Society
The Mansion House
Bonehill
Chiswell Green Lane
Chiswell Green
St Albans
AL2 3NS
www.royensoc.co.uk

By the
FSC Publications
Unit C1
Stafford Park 15
Telford
TF3 3BB
www.field-studies-council.org

ISBN: 978-0-901546-94-4

# Contents

| | |
|---|---|
| **Acknowledgements** | iv |
| **Introduction** | 1 |
| **Life cycles and biology** | 2 |
| **Morphology** | 3 |
| **Collecting and preserving** | 9 |
| **Dichotomous keys and other approaches** | 12 |
| **Notes on species accounts** | 13 |
| **Higher classification** | 14 |
| **Check list of British species** | 15 |
| **Key to families** | 23 |
| Family Rhyacophilidae | 27 |
| Family Glossosomatidae | 30 |
| Family Hydroptilidae | 36 |
| Family Philopotamidae | 54 |
| Family Ecnomidae | 60 |
| Family Polycentropodidae | 61 |
| Family Psychomyiidae | 72 |
| Family Hydropsychidae | 81 |
| Family Phryganeidae | 90 |
| Family Brachycentridae | 102 |
| Family Goeridae | 104 |
| Family Lepidostomatidae | 107 |
| Family Apataniidae | 110 |
| Family Limnephilidae | 112 |
| Family Sericostomatidae | 150 |
| Family Beraeidae | 152 |
| Family Odontoceridae | 156 |
| Family Molannidae | 157 |
| Family Leptoceridae | 158 |
| **References** | 181 |
| **Index** | 187 |

# Acknowledgements

Most of the drawings of wing venation used in this handbook were originally prepared by the late D.E. Kimmins for Mosely's (1939) book. The drawings of Hydroptilidae are from Marshall (1978, 1979), and the figures of female genitalia of *Tinodes* are from Fisher (1977). The photos of living caddisflies are by Emma Ross except where indicated, as are most of the genitalia drawings; all photos of preserved specimens and microscope slide preparations are by Peter Barnard.

We thank Dr Ian Wallace (National Museums of Liverpool) for much advice, information and encouragement during the production of this handbook; he also generously provided distribution notes which were compiled from the Trichoptera database at the Biological Records Centre in preparation for a distribution atlas. Dr Jim O'Connor (National Museum of Ireland, Dublin) provided helpful information on Irish specimens and their distribution. David Pryce kindly sent us records and specimens of some rare species. Stuart Crofts gave us much information on the capture of *Synagapetus dubitans* in advance of his published note, and also provided the photo of a live *Brachycentrus subnubilus*. Andrew Godfrey sent useful comments on the populations of *Sericostoma* in Britain and elsewhere.

We also thank Dr Malcolm Scoble, former Keeper of Entomology at the Natural History Museum in London, for continued support and encouragement after both authors had left the museum.

# Introduction

There have been two major works covering the identification of adult caddisflies in the British Isles (Mosely, 1939; Macan, 1973) but the group still has the reputation of being difficult to identify except by an expert. No doubt this is partly due to the fact that close examination of the genitalia is almost essential for the correct determination of most species, but even the beginner to the group can make good progress by studying other distinguishing characters. In a previous publication (Barnard & Ross, 2008) we have shown that some of the more common species can be recognised by their wing patterns, a trend that was started by Mosely and followed by Macan in their respective handbooks, though using only monochrome photographs in both cases. This current work continues the use of colour photographs, but here we have illustrated pinned 'museum' specimens of most of the 199 British and Irish species, combined with a range of living specimens.

Interest in this group of freshwater insects has increased greatly in Britain over recent years. Inevitably there is a great fascination in the larvae of caddisflies, with their well-known case-building habits and their abundance in many kinds of freshwater. Larvae can now be easily identified using the keys of Edington & Hildrew (1995) for the free-living groups, and Wallace, Wallace & Philipson (2003) for the case-building groups, both following on the pioneering work of Hickin (1967).

Adult caddisflies, on the other hand, are often dismissed in general entomology books as being moth-like and uniformly brownish in colour, a misconception we hope to rectify here. As Mosely (1921) put it, "the Trichoptera are not all studies in sober browns and greys". Several of the common adults are well known to fly-fishermen, who imitate many species with their artificial flies, and they are also frequent visitors to the light-traps of moth recorders; however, the absence of a modern and accessible key has made detailed identification difficult. This handbook should encourage entomologists to take a closer look at the specimens they might otherwise have rejected as being unidentifiable.

It should be noted that the term 'British' is deliberately used rather loosely here; it generally refers to Great Britain (i.e. England, Scotland and Wales) because Northern Ireland is generally covered in the Irish literature. Irish distributions are not described in detail but their complete exclusion would mean that the three species found only in Ireland and not Great Britain would have to be excluded. Traditionally all the species in the British Isles have been treated together.

The pioneering work on the European caddisfly fauna was McLachlan's (1874-1880) great revisional synopsis, which included 474 species; the latest work to cover approximately the same region is the second edition of Malicky's (2004) *Atlas* which includes over 1400 species, a figure that continues to rise every year. Over the same period the number of British species has risen from 136 in 1870 to 199 today, the smaller increase no doubt reflecting the well-worked fauna of the British Isles. Mosely's (1939) handbook relied heavily on McLachlan's 19th century work, even to the extent of reproducing many of his illustrations of genitalia, with only partial success. Macan's (1973) handbook broke new ground with entirely new keys and illustrations, though with several difficulties that have deterred all but the most determined novices to the group. As well as these handbooks, checklists of the British species were published by Kimmins (1966) and Barnard (1985) and the various changes in nomenclature are noted in the checklist given below as well as in the appropriate places in the main text. Most of the current changes reflect work that has been done across Europe, particularly with complex species groups in *Rhyacophila*, *Hydropsyche* and so on, and these are discussed in papers such as Malicky (2005). Recent distributions of species at country level throughout Europe can be found on the Fauna Europaea website (http://www.faunaeur.org) and there is a regularly updated checklist of the world Trichoptera (http://entweb.clemson.edu/database/trichopt) which currently shows a world total of well over 12,000 species.

There is a very active worldwide community of trichopterologists with an annual journal *Braueria* and a regular series of international symposia, which began in 1974. The published proceedings of these symposia form a useful summary of current work on Trichoptera, and many are listed in the references at the end of this handbook. Interest in the study of caddisflies as part of a wider awareness of freshwater insects in Britain is promoted by The Riverfly Partnership (http://www.riverflies.org/index) and a recording scheme for the British caddis is run by Dr Ian Wallace (http://www.brc.ac.uk/schemes/RRS/trichoptera.htm).

# Life-cycles and biology

The biology of the aquatic stages of the Trichoptera is well documented, and many references will be found in the two handbooks to larvae (Edington & Hildrew, 1995; Wallace *et al.*, 2003) but there are many aspects of adult biology that need to be understood in order to correctly identify the significance of certain records of Trichoptera species. As with most aquatic insects such as Ephemeroptera and Plecoptera the adult phase can be very short-lived, just sufficient for mating, dispersal and oviposition, but in some Limnephilidae the adult phase lasts for some months, as detailed below. Flight periods are quoted for most of the British species, but these can vary considerably in different parts of Britain, generally because of average temperature differences at different latitudes. There are several papers documenting these changes, as well as seasonal flight activity in general, by Crichton and co-authors (Crichton, 1971; 1987; Crichton, Fisher & Woiwod, 1978; Crichton & Fisher, 1981; 1982).

Some species have more than one peak of flight activity, which may represent at least two generations per year, but such bivoltinism may only occur in the south of Britain. Many species of Limnephilidae undergo a reproductive diapause in the summer months, apparently in response to the drying-up of the temporary water bodies in which their larvae develop. Again, this diapause may not occur in the cooler temperatures in northern Scotland. Marshall (1978) discusses flight periods of Hydroptilidae, and Cooling (1982) describes the correspondence of adult and larval records on several rivers in southern England.

The mating behaviour of caddisflies is extremely complex, and there is clearly a great deal more to be discovered on this subject. Those species with a summer diapause emerge in the spring and usually mate soon afterwards but with no subsequent oviposition; after several weeks of inactivity they often mate again before the females lay their eggs in the autumn. The reason for this double mating is not clear, and even those species without a diapause still show complex behaviours that are poorly understood. Some general aspects of mating behaviour are discussed by Hoffmann (1999) and Syrnikov *et al.* (2005). Multiple mating seems to be common in many species, and such polyandry is described by Petersson (1991). Denis (1981) showed how ovarian maturation is controlled by photoperiod in both the larval and adult phases, and Malicky (1991) discusses some of the wider implications of the adult diapause.

Caddisflies produce a variety of pheromones, some important in attracting potential mates, some used to aggregate swarms of males, and others with a defensive function. The main pheromone glands are situated on the abdomen, but there are numerous other structures on the head and wings of the males that are also assumed to be sites of pheromone production. Ivanov & Löfstedt (1999) have analysed the chemical components of pheromones in Trichoptera, and Ivanov & Melnitsky (2002) describe the morphology of the abdominal glands. Some of these pheromones are very acrid to the human senses, and it is best to avoid using mouth-operated pooters for collecting the larger caddisflies. There are other methods of general communication between adult Trichoptera, including wing vibration and drumming of the body or wings on the substrate, as well as waving of wings and antennae: these are all reviewed by Ivanov (1997).

The flight behaviour of males and females is often different; the males of many species form dense swarms that attract receptive females, a behaviour seen most noticeably in the day-flying Leptoceridae (known to anglers as the "long-horns"). During the summer months these swarms are seen in the late afternoon on the banks of lakes or slow-flowing rivers, usually centred on a fixed visual marker such as a tree (or even an unwitting human observer!). The males fly in regular patterns within the swarm, and their long pale antennae form a distinctive visual effect, reminiscent of the lepidopteran family Adelidae. In other families mating may occur on nearby vegetation. The flight behaviour of ovipositing females depends on whether the eggs are laid above or in the water, and it also differs between groups frequenting running or still water. Some of these sexual differences in flight behaviour were discussed by Solem & Bongard (1987).

Oviposition is a critical phase in the life-cycle of most insects, and caddisflies have evolved a variety of means to ensure that eggs are safely deposited in the correct environment in order to maximise the chances of survival of the hatching larvae. The females of those species inhabiting still water can often simply drop their eggs near the water margins or near emergent vegetation; others will walk below the surface to attach them directly to submerged plants or to stones at the bottom. Even in fast-flowing water, females of groups like *Brachycentrus* will walk below the surface to attach their egg-mass to a fixed substrate, a behaviour which clearly prevents the eggs from being washed rapidly downstream (Barnard, 1978). Most of the species living in fast water will show some upstream flight activity before laying their eggs in order to counteract the tendency for their eggs to be washed too far downsteam. In the family Limnephilidae the eggs are usually laid in a gelatinous mass, often above the water level, and eggs are stimulated to hatch either by rising water levels or by rain. This adaptation to living in temporary waters is seen most dramatically in *Glyphotaelius pellucidus*, a common inhabitant of temporary woodland pools. In this species the egg-masses are laid on leaves of trees overhanging the sites of the dried-up pools; the autumn rains cause the larvae to hatch and drop into the pools gradually filling up below them (Crichton, 1987). Many people have been puzzled to find such egg-masses high and dry in woodland, and are surprised to learn that they belong to an aquatic insect.

# Morphology

It is essential to understand the basic morphology of adult caddis before any identification can be successful; fortunately the classification of the group currently relies on relatively few morphological characters that are not too difficult for the novice to learn. First of all, one must be able to distinguish a caddisfly from any other insect. The nearest neighbours of the

Trichoptera are the Lepidoptera (butterflies and moths) and even experienced entomologists sometimes confuse the smaller members of these two insect orders, which in evolutionary terms are considered as sister-groups. As well as a similar general appearance they have similar wing venation, both groups have long tibial spurs, and some smaller moths do not have the coiled proboscis so characteristic of butterflies. The easiest way to separate the two groups is by studying the covering of the wings; Trichoptera have hairs on their wings (*trichos* in ancient Greek means a hair) and the Lepidoptera have flattened scales (from *lepidos*, a scale) (Figs 1, 2). Some male caddisflies have numbers of flattened scales, but these are usually confined to distinct areas of the wing, and some moths have tufts of hairs (or even extensive mixtures of hairs and scales, as in the Psychidae), but in general the distinction is fairly clear. Also, moths often have scales over the rest of the body including the legs and even the external genitalia, a feature not seen in caddisflies. The micro-structure of caddisfly wing hairs has been studied by Moss & Gibbs (2000).

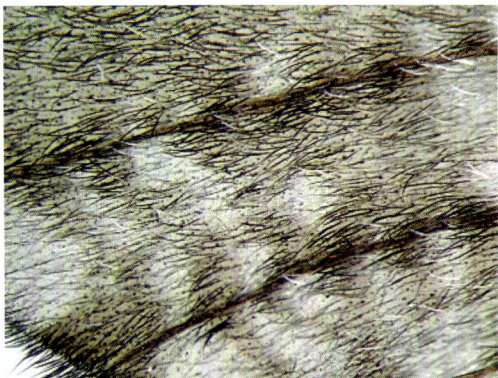

Figure 1. Magnified view of hairs on caddisfly wing          Figure 2. Magnified view of scales on butterfly wing

Distinguishing caddisflies from other groups of aquatic insects is relatively easy. As can be seen from the photos of living insects in this handbook, most caddisflies sit with the wings folded over their back like a roof or ridge-tent, the long antennae are often held out straight in front of the head, and there are no long 'tails' at the end of the body. All these features distinguish them from other common groups of freshwater insects. Ephemeroptera (upwinged flies, or mayflies) hold their wings upright over the body at rest and have two or three long 'tails'; Odonata (dragonflies and damselflies) hold their wings out sideways or upright, and have very short antennae; Plecoptera (stoneflies) hold their wings flat across the body (or sometimes wrapped round it) and usually have two 'tails'. None of these other freshwater groups have hairy wings like caddisflies. Some other groups of insects such as lacewings (Neuroptera) include aquatic species that may have hairs along the veins of the wings but never all over the membrane.

Beginning with the head, the most prominent appendages are the antennae, which in most caddisflies are simple and composed of numerous similar-sized segments, and with a total length approximately the same as the fore-wing length. In those species with exceptionally long antennae, principally in the family Leptoceridae, the extra length is created by an elongation of each segment, rather than by increasing the number of segments. Some groups show unique features: *Odontocerum albicorne* has short tufts of hairs on each antennal segment, giving a toothed appearance, and the species of *Hydropsyche* have a dark curved ridge on each segment that create the impression of a helical line running along the whole antenna. The two basal segments of the antennae, the scape and pedicel, sometimes show extreme modifications in the male, which are described in the appropriate species accounts.

From a functional point of view, the mouthparts of adult Trichoptera are rather simple; the mandibles and maxillae are much reduced and all the feeding activity is carried out by a large protrusible haustellum. This has a highly complex grooved surface with pectinate hairs, used to lap up liquids (Crichton, 1991). The most distinctive components of the mouthparts are the palps, and these are widely used as taxonomic characters. The labial palps are always three-segmented and unmodified but the maxillary palps, which are anterior to the labial palps, can vary considerably. In the female the maxillary palps are always five-segmented (Fig. 3) but in the male the number of segments ranges from two or three to five, and they are often highly modified. The most extreme forms are seen in males of the Sericostomatidae, Goeridae and Lepidostomatidae, where the segments are fused and swollen into such bizarre shapes that the original segmentation cannot be recognised (Fig. 4). In male Limnephilidae and Brachycentridae the maxillary palps are three-segmented (Fig. 5), and in the male Phryganeidae they are four-segmented, usually little modified from the typical female form. The relative lengths of the maxillary palp segments were widely used taxonomically by authors such as Mosely (1939) but it is not easy to distinguish these subtleties in dried pinned specimens, though it is easier in fluid-preserved ones.

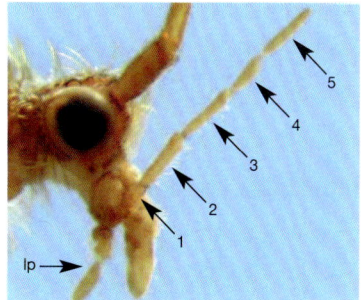

Figure 3. Lateral view of female head showing 5-segmented maxillary palp; lp = labial palp

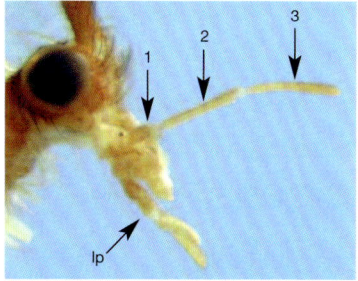

Figure 5. Lateral view of male head showing 3-segmented maxillary palp; lp = labial palp

Figure 4. Modified maxillary palps in male *Sericostoma*

The large compound eyes are relatively uniform in appearance throughout the whole order, although there can be some sexual dimorphism in size, as in *Hydropsyche exocellata*, where the male eyes are much larger. The presence or absence of ocelli are important in the classification of the group, and it is essential to be able to distinguish the ocelli from other structures on the head, not always easy when the head has a dense covering of hairs. When present, the three ocelli are arranged in a triangle, with the two dorsal ones between the compound eyes, and the anterior median one near the bases of the antennae (Fig. 6). In the larger species such as the Phryganeidae, the ocelli are easy to see, because they have a

Figure 6. Dorsal view of ocelli on head

whitish translucent appearance under a bright light, but in smaller groups, especially the Hydroptilidae, it takes some practice to distinguish them from the various 'warts' or 'cushions' also found on top of the head. These other structures, usually with tufts of long hairs, can be characteristic for certain taxonomic groups, but they are rarely used in practical identifications. A recent attempt has been made to devise a system of nomenclature for these structures (Oláh & Johanson, 2007).

Similar 'cushions' of hairs are found on the thorax, and their taxonomic value was recognised by authors such as Ross (1944). Macan (1973) tried to make further use of such characters, but there are several practical difficulties. In conventionally pinned museum specimens the pin often obscures these thoracic structures, though this difficulty does not arise with fluid-preserved material. Unfortunately the characters are not always consistent, with sexual dimorphism seen in the Lepidostomatidae and no consistent pattern at all within the Beraeidae, so these characters are not used in this handbook.

The legs of caddisflies show little structural modification, except that the middle legs are broadened in the females of some genera such as *Glossosoma*; this seems to be an aid to swimming under water while the female is searching out an oviposition site. However, the number and arrangement of spurs on each

Figure 7. Tibial spurs (pale, arrowed) and spines (dark)

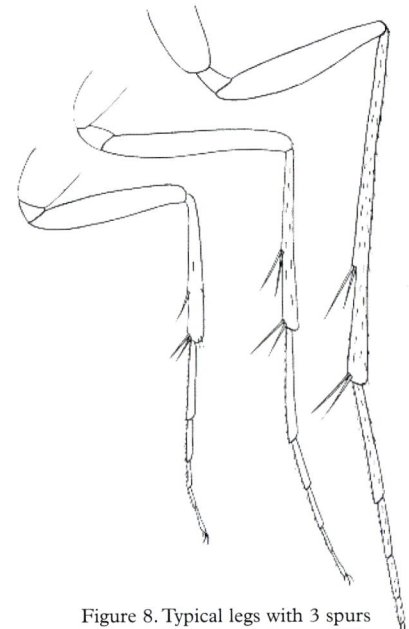

Figure 8. Typical legs with 3 spurs on fore tibia and 4 each on middle and hind tibia

of the tibiae are widely used in identification. First it is important to distinguish between spurs and other hairs or spines on the legs, as the latter are common in the Phryganeidae and Limnephilidae. Spurs are generally longer than spines (Fig. 7) and are articulated at the base, though they are movable only in fresh specimens or those preserved in fluid. Spurs are usually pale brown or yellowish, while spines are dark brown or black. In addition, spurs are found only at the distal end of the tibia or up to half-way along it. There are never more than two spurs at one position, so the absolute maximum to be found on one leg is four, with a pair at the apex and another pair somewhere in the middle of the tibia. The number of spurs on each leg – front, middle and hind – gives the so-called spur formula. Thus the example in Fig. 8 has the formula 3.4.4, because there are three spurs on the front leg, four on the middle, and four on the hind. It is important to check the numbers on both sides of the body, because spurs can be lost, giving an artificially low count. It is also worth remembering that there are never more spurs in the middle of the tibia than at the end. So if a leg appears to have a pair of spurs in the middle and only one at the apex, then one of the apical spurs is missing, and the total number should be four, not three. The number of spurs on the front leg varies between none and three, and on the middle and hind legs it can be from two to four. Another pitfall for the unwary is that the spurs can be very small and almost hidden by hairs. In some species the spurs on the front leg may be so much smaller than those on the other legs that they are easily overlooked, but mistakes in counting the spurs will diminish with experience.

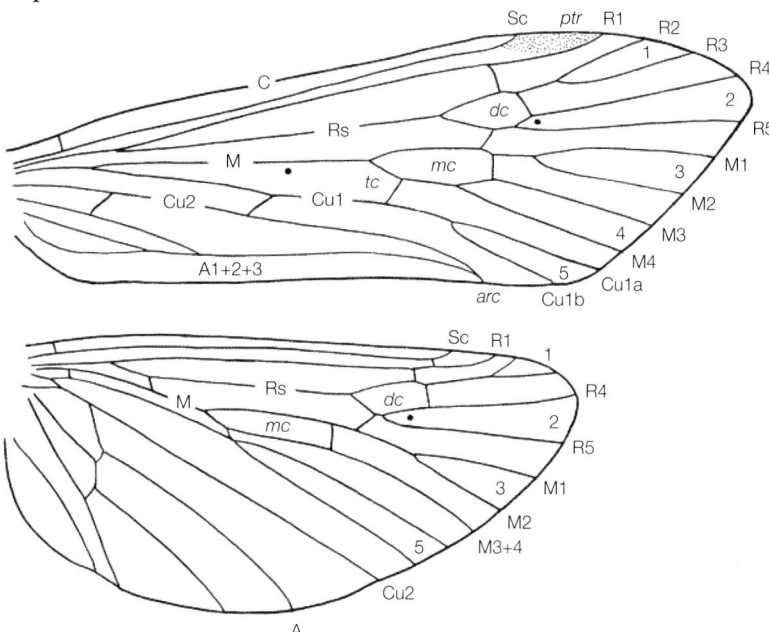

Figure 9. Generalised wing venation (for abbreviations see text)

The wings are the most obvious appendages of the thorax, and their general shape, size, colour and especially venation are the most widely used characters for identifying caddisflies, at least to generic level. The fore wings are often much thicker than the delicate membranous hind wings, and any coloured patterns are usually on the wing membrane itself. Thus the pattern can still be seen in specimens denuded of hairs, unlike the Lepidoptera where the pattern is usually formed by the wing-scales. The generalised pattern of venation is shown in Fig. 9, with the main veins labelled in the conventional way. In the fore wing the front margin is formed by the costa (C), closely paralleled by the subcosta

(Sc). The radial vein divides near the base of the wing into the radius (R1) and the radial sector (Rs) which further divides into as many as four branches (R2-R5). Between Sc and R1 at the wing margin is the darkened pterostigma (*ptr*). The median vein (M) also divides into up to four branches (M1-M4). The remaining cubital (Cu1 and Cu2) and anal veins (A1-A3) are less divided, and sometimes fuse in various ways. There are various 'landmark' features, defined by the venation. Where vein Rs first divides, the discoidal (or discal) cell (*dc*) is formed, and where this cell is closed off by a crossvein between R2+3 and R4+5 the discoidal cell is described as closed (or present). When that crossvein is missing, the discoidal cell is considered as open (or absent). A similar cell is formed where M first divides and this median cell (*mc*) can again be either closed (present) or open (absent). There is often a pale, hairless spot termed the thyridium near the main fork of M and the large cell between the bases of M and Cu1 is therefore called the thyridial cell (*tc*). Another similar pale spot is often found where the fused anal veins meet the hind margin, and this is called the arculus (*arc*). Before a rationalised system of naming wing-veins had been developed, the apical forks of the wings had already been given arbitrary numbers, starting from the front of the wing. Despite some attempts to rename these forks (e.g. Kimmins, 1956) the older system is still in regular use (e.g. Schmid, 1998). These forks are labelled 1 to 5 in Fig. 9, and it should be noted that fork 2 (formed by R4 and R5) can be recognised by the dark 'corneous spot' near its base. Forks with a separate stem (such as forks 1 and 3 in Fig. 9) are said to be stalked; those without such a stem (such as fork 4 in Fig. 9) are described as sessile. The series of crossveins extending from front to back of the fore wing at about the levels of the discoidal and median cells is called the anastomosis. In some families such as the Leptoceridae this can be seen as a distinctive straight line, and in genera like *Mystacides* this anastomosis forms a line of folding across the wing. The hind wing venation is similar to that of the fore wing, the main difference being that the anal veins are more numerous and divergent in the hind wing, thus supporting the wide anal lobe seen especially in families like the Limnephilidae. Vein M3+4 never divides, so fork 4 is always absent in the hind wing. Sexual dimorphism in wing venation is quite common, and this is discussed under the corresponding species accounts.

The abdomen of caddisflies has several unusual features, including the almost universal paired openings of the pheromone gland between segments four and five. Many species have ventral processes on various segments in either or both sexes, which may have importance during copulation, but they have not yet been studied systematically. A few groups have long filaments on the abdomen arising from the fourth or fifth segment (Polycentropodidae, and *Diplectrona* in the Hydropsychidae). By far the most important structures on the abdomen are the genitalia, which show interspecific differences in both sexes in nearly all species; their study is therefore vital to a correct identification.

The male genitalia show a huge range of diverse structures, and their correct homology across the whole of the Trichoptera is extremely difficult to establish. For this reason, some of the structures still retain the arbitrary names given by early workers, and it remains to be seen whether attempts to homologise all the components in an evolutionary framework are successful (Ivanov, 2005). The eighth segment of the male abdomen is usually the last unmodified one; on the ninth segment the ventral sternite and dorsal tergite are usually fused into a more-or-less complete ring (IX in Fig. 10). The main processes attached to this are a pair of large claspers (clasp. in Fig. 10), or inferior appendages often two-segmented, which hold the female during copulation. There is a small but prominent segment ten, termed the dorsal plate in earlier works such as Mosely (1939). This bears the dorsally situated superior appendages (sup. app.) and below them the intermediate appendages (int.

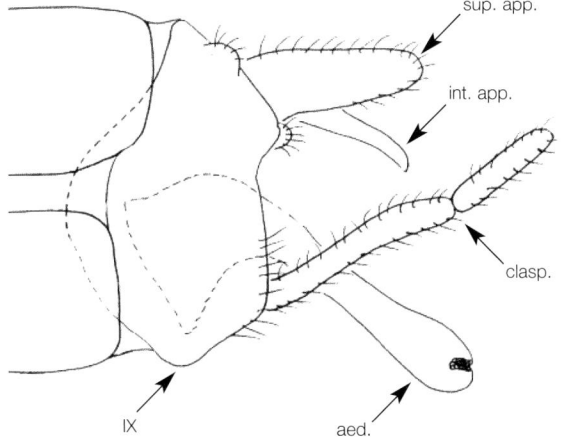
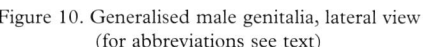
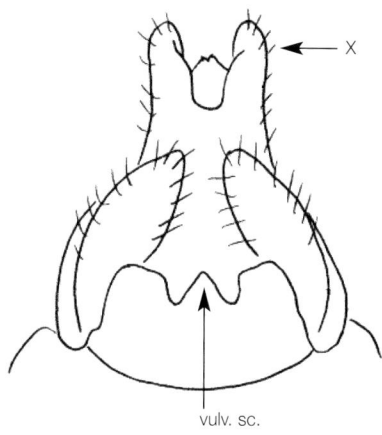

Figure 10. Generalised male genitalia, lateral view (for abbreviations see text)

Figure 11. Generalised female genitalia, ventral view (for abbreviations see text)

app.), called the mid appendages by Macan (1973). All three pairs of appendages can vary greatly in size and shape, and they are not always present; the terms superior and inferior refer to their position on the segment, and not to their size or importance. The intromittent organ, through which the sperm duct passes, is even more problematical to homologise, but there is usually a central apparatus termed the aedeagus (aed.) phallus or penis, and a pair of lateral structures called parameres. These organs may be introverted inside the abdomen, or extruded externally, and the resulting differences in their appearance can be very confusing. For this reason they are not used for identification in this work.

The female genitalia are superficially simpler than those of the male, but their diversity is just as great. In this sex (Fig. 11) the eighth sternite (or subgenital plate) often bears a central vulvar scale (vulv. sc.), which may be divided into three lobes. The ventral part of segment nine is often absent, and segment ten (X in Fig. 11) is usually fairly prominent. It often bears few appendages or none, but in some groups there are up to three pairs, which can make the distinction between the sexes less obvious than one might wish. There are often complex internal structures that become apparent when genitalia preparations are made.

Gynandromorphs (intersexes) and other genital deformities are not particularly rare in Trichoptera, so it is unwise to assume that any odd-looking specimen is a new species! Wing venation can also be rather variable, often within local populations (e.g. Fox, 1957).

## Collecting and preserving

Many species, though not all, are attracted to light-traps of the types used by moth recorders. The attraction of the light, whether UV or actinic, can be very different between closely related species and can also vary markedly between the sexes of the same species. Humid, or even rainy, nights can be very productive, and caddis are one of the first insect groups to appear after dark, with a peak of activity an hour or two after sunset. Therefore the caddisfly collector can usually retire for the night much earlier than his lepidopteran colleagues! For the groups not attracted to light direct collecting methods must be used; some day-flying species can be caught with a standard insect net, and some can be swept from low vegetation overhanging the water. Beating trees near the waterside can be very productive but it is usually best to beat into a net rather than a beating-tray because many species are very lively when disturbed and will quickly fly away. Some plants will be found more productive than others; ash 'keys' often harbour diapausing adult Limnephilidae, while stream-dwelling species may shelter in ferns, but more information is needed about the resting places of many species. Caddisflies are also frequently caught in Malaise traps and other interception traps in suitable habitats. Specimens that are to be preserved in alcohol can be killed by collecting directly into tubes of 80% alcohol (see details below), but those to be preserved dry can be killed with ethyl acetate vapour in a standard insect killing jar, while some collectors simply 'tube' specimens and kill them later in a freezer.

Collecting must of course be carried out responsibly, by following the code for collecting (Invertebrate Link, 2002). A personal collection is useful for learning the identification of a group and the retention of voucher specimens can be important for future reference. Very few of the rarer British species can be recognised unambiguously in the field; *Hagenella clathrata* may be the one exception. However, it is worth noting that some of the larger specimens such as Limnephilidae and Rhyacophilidae can be identified live in the field; if the wings are held together the insect will often curl its abdomen down, exposing the genitalia. The experienced worker can then use a hand-lens to identify the specimen which is then released unharmed.

Traditionally caddisflies have been pinned and 'set' with the wings spread, like butterflies and moths, and this is the way that most older museum specimens will be found. This method has the advantage that the wing patterns and venation can be seen relatively easily, although the patterns may fade fairly rapidly, as can be seen in the photos of museum specimens used in this handbook. The disadvantage of this method is that the genitalia cannot be examined without removing them and softening them with chemical means; removing abdomens often results in the specimen breaking across the thorax. The maxillary palps shrivel when they dry, and the tibial spurs may also be hard to see. There are therefore advantages in storing specimens in fluid such as alcohol, which keeps all the appendages looking natural, and maintains some flexibility so that parts can be manipulated to examine them thoroughly. The disadvantages are that the wing patterns are rather obscured by immersion in liquid and the folded hind wings are especially difficult to examine. Comparing such specimens is not a rapid process, when each has to be removed from a storage tube. Clearly there are advantages in both methods of storage, a fact noted by Mosely (1939), but keeping series of specimens in two different systems may be inconvenient. It can be difficult for the amateur worker to obtain ethyl alcohol (ethanol) in sufficient strength for preserving collections; commercially available alcohols such as methylated spirits are not suitable because they turn cloudy when mixed with water. However, good results can be obtained by using the readily available isopropyl alcohol (isopropanol) which should be diluted to about 80% with distilled water, i.e. four parts by volume of isopropanol to one of distilled water.

Even with alcohol preserved specimens it may be necessary to 'clear' the genitalia so that the internal structures can be seen, and most of the drawings in this handbook are made from such preparations. The reason for this can be seen in the illustrations showing the differences between three specimens of the same species (Figs 12-14). In dried specimens (Fig. 12) the main outline of the genitalia is visible, but many structures are contracted and sometimes even inverted, though an expert can still recognise many species in this state. Specimens preserved in alcohol (Fig. 13) retain a more natural appearance with less distortion but the internal structures are still invisible. Cleared specimens (Fig. 14) not only show these

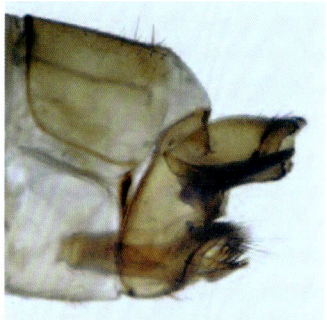

Figure 12. Limnephilus flavicornis, female genitalia of dry specimen

Figure 13. L. flavicornis, female genitalia preserved in alcohol

Figure 14. L. flavicornis, female genitalia cleared in KOH

internal features but they also clarify the relationships between the main components in a three-dimensional way. Whether or not clearing is necessary for any particular specimen will become obvious with experience but in some groups such as the Hydroptilidae and females of *Hydropsyche* it is nearly always essential.

Abdomens are cleared by soaking in dilute (5% or 10%) solutions of potassium hydroxide (KOH) or sodium hydroxide (NaOH), the latter being readily available as the drain cleaner, caustic soda. Both are highly caustic and must be used with great caution. Clearing small specimens such as hydroptilids may take only an hour or two, while large caddisflies will need several hours, and are often left to soak overnight. The unwanted abdominal contents (gut, muscle and fat-bodies) can then be removed with fine needles under the microscope; the abdomen is next washed in distilled water several times to remove the caustic hydroxide, and then it can be preserved, either in a small vial of alcohol with the original specimen, or in a micro-vial of glycerine attached to the pin of a dry specimen. It is, of course, essential to ensure the correct association of the abdomen with the rest of the insect.

Collectors of large numbers of insects, as may be inevitable in surveys based on light-trap catches for example, may want to retain only voucher specimens in a permanent collection. In such cases caddisflies can be stored dry and unset in paper packets, and batches can later be cleared in bulk. This has the disadvantage that the wings can be damaged by prolonged exposure to the caustic solution but at least large numbers of specimens can be identified quickly; one has to accept that the single specimen of that rare species may not be an attractive example of its kind!

The standard methods of maintaining insect collections can be found in several entomological textbooks. Whatever method of storage is adopted, the importance of correct labelling cannot be over-emphasised, as one can never predict what the future importance of a collection might be. There is nothing more frustrating to a researcher than finding incomplete or ambiguous data that cannot be reconstructed.

# Dichotomous keys and other approaches

It is now generally recognised that dichotomous keys are not always easy for beginners, or even experienced workers, to use for the identification of organisms. Whenever a difficulty is encountered the user has little confidence in being on the right track throughout the rest of the key, and it is dispiriting to arrive at a clearly wrong answer after a long session with impenetrably descriptive key couplets. A further complication with caddisflies is that specific differences are usually seen only in the genitalia, which often necessitates different keys for each sex. This leads to long keys in which each key couplet has a lengthy description of complex genitalia, followed by "not as above" in the second half, a scenario that deters all but the most determined user. In recent years this recognition of the limitations of a traditional key have led some authors to abandon keys altogether and to produce a pictorial guide with minimal text (Malicky, 1984a). This so-called 'atlas' approach provides all relevant illustrations for each taxon; similar species are grouped together for easy comparison, and minimal text is included on critical characters, distribution and so on. However, we feel that a simple 'atlas' to the 199 British species, along the lines of Malicky's (2004) work on the European fauna, could give rise to other difficulties caused by searching for species in an entirely wrong section of the book. So here we have adopted a combined approach: there is a key to families, and within each family there are keys to genera or generic groups. So having provided a framework of the higher classification with conventional keys, the groups of species can be compared directly with each other and this is where the 'atlas' style should prove its worth.

If the user finds difficulty with the conventional keys, then the handbook can still be used as a simple atlas of species, though we would suggest that flicking through all the pictures is not a very efficient way of using the book! Every user will develop a preferred way of using a combination of keys and pictures, and this will change as experience is developed. However, we recommend that the text for each species should be read carefully before an identification is confirmed. Problems of identification for all species are listed, and comparisons with similar species are emphasised. Even the size range of a species is important: we have seen many specimens, particularly females, placed by a novice in the wrong family, when a simple checking of the recorded size or other easily observed characters would have made the error obvious. Details of wing venation are not always easy to see, and even some experts try to avoid using them, but a careful examination will often help to confirm that one is looking in the right group. For this reason, the main venational characters are listed under each family and genus.

Unique diagnostic characters are indicated with arrows on the appropriate figures, but we would always urge that the corresponding text should be read to confirm that a closely related species has not been overlooked. Other apparent differences between species are not necessarily reliable if they are not arrowed.

Note that species are listed by natural groupings within each family and genus, and not in alphabetical order. This is so that similar species can be compared more easily, and is consistent with the atlas approach. Mosely (1939) used a similar principle, though without specifically stating so.

An alphabetical index to all taxa is included at the back.

# Notes on species accounts

Under each species will be found the following information (reduced information is given for the three species found only in Ireland):

• The correct scientific name with taxonomic notes where this differs from the last published checklist (Barnard, 1985), including notes where the name differs from that used by Mosely (1939) and Macan (1973)

• Common names used by anglers are noted where known, and any current conservation status is listed; some species have recently been given BAP status (the subject of Biodiversity Action Plans; see http://ukbars.defra.gov.uk/)

• A brief description of the wing pattern, together with photos of pinned and living specimens and in many cases a diagram of the wing venation

• The fore wing length with its usual range is given, though exceptionally large or small specimens may be found

• A summary of the known distribution in Great Britain, generously provided by Dr I.D. Wallace from the Caddis Recording Scheme data, updating the information in his earlier review (Wallace, 1991); followed by a brief indication of the European distribution. Detailed information on Irish species can be found in O'Connor (2010) and the references therein

• The main flight period (with less common months in brackets) and common habitats; more details on larval habitats are in Edington & Hildrew (1995) and Wallace *et al.* (2003), though it must be noted that some adult caddisflies can be captured far from their breeding sites because of dispersal flights

• The genitalia of both sexes are figured, with arrows indicating critical characters to separate species from their nearest neighbours. In most drawings setae are omitted unless they are very prominent; in dorsal or ventral views they are generally omitted on one side to show morphological features that would otherwise be hidden. As explained above, the male aedeagus is often omitted from the figures

In addition, the number of European species in each genus is indicated to show how the British fauna compares, though the continental totals are only an approximation because of imprecise information for some areas of Europe.

The standard symbols of ♂ for male and ♀ for female are used throughout.

# Higher classification

The higher classification of the Trichoptera is still hotly debated, with devotees of molecular phylogenetic methods finding difficulties in agreeing with those who put more weight on conventional morphological, ecological and behavioural studies; see Dreesmann & Wichard (2002) and Holzenthal *et al.* (2007) for some recent discussions. A relatively conservative approach has been adopted here, on the grounds that any ultra-modern system is likely to change whenever new molecular data are discovered.

Currently three suborders of the Trichoptera are generally recognised, containing the following families (those not found in Britain are excluded):

Suborder Spicipalpia (primitive, cocoon-making caddis)
    Rhyacophilidae
    Glossosomatidae
    Hydroptilidae

Suborder Annulipalpia (caseless, retreat-making caddis)
    Philopotamidae
    Ecnomidae
    Polycentropodidae
    Psychomyiidae
    Hydropsychidae

Suborder Integripalpia (true case-building caddis)
    Phryganeidae
    Brachycentridae
    Goeridae
    Lepidostomatidae
    Apataniidae
    Limnephilidae
    Sericostomatidae
    Beraeidae
    Odontoceridae
    Molannidae
    Leptoceridae

As can be seen from the brief descriptions of each suborder these are based more on larval characters than adult ones, although the actual names of these higher groups are derived from adult characters, as is traditional in much taxonomic work. It is generally agreed that only the Annulipalpia and Integripalpia are monophyletic.

# Check list of British species

The last complete check-list of the British and Irish Trichoptera was published over twenty-five years ago (Barnard, 1985). There has been one recent addition and just one deletion to the species list during that period but there have been several changes in nomenclature, including the recognition that one species (*Ecclisopteryx guttulata*) was incorrectly identified in the British Isles. Perhaps the most striking change is the appearance of the family Apataniidae, previously considered as a subgroup of the Limnephilidae. Notes on all these changes will be found under the appropriate species accounts. Of the 199 species currently on the list, three (*Apatania auricula*, *Limnephilus fuscinervis* and *Tinodes maculicornis*) are known only from Ireland, not Great Britain, even though all three are found elsewhere in Europe. Thus there are 196 species currently known from Great Britain, with 150 in Ireland (Ashe *et al.*, 1998; O'Connor, 2010). The handbook will also work for all species currently known from the Channel Islands, although the exact status of some of the forms of *Philopotamus* found only in those islands is uncertain.

The recent discovery of *Synagapetus dubitans* shows that the collector must always be alert to the possibility of continental species being discovered in Britain, and there is also the possibility of exotic species being accidentally introduced along with imported aquatic plants. It is highly likely that some of the hydroptilids discovered by Mosely in the early 20th century, such as *Orthotrichia tragetti*, were introduced in such a way.

It should be noted that in Macan's (1973) handbook *Oligotricha* was consistently mis-spelt "*Oligotrichia*", *Athripsodes* as "*Arthripsodes*" and *Neureclipsis* as "*Neureclepsis*".

Subgroups are given only for the largest family, the Limnephilidae, to assist with understanding the generic relationships although on a worldwide scale many other families are divided into subfamilies and tribes. In every family the names are listed alphabetically within each higher grouping.

**SUBORDER SPICIPALPIA**
**Superfamily Rhyacophiloidea**
**Family RHYACOPHILIDAE**
      **RHYACOPHILA** Pictet, 1834
         dorsalis (Curtis, 1834)
         fasciata Hagen 1859
            = *septentrionis* McLachlan, 1865; synonym
         munda McLachlan, 1862
         obliterata McLachlan, 1863

**Superfamily Glossosomatoidea**
**Family GLOSSOSOMATIDAE**
      **AGAPETUS** Curtis, 1834
         delicatulus McLachlan, 1884
         fuscipes Curtis, 1834
         ochripes Curtis, 1834
            = *comatus* Pictet, 1834; synonym
      **GLOSSOSOMA** Curtis, 1834
         boltoni Curtis, 1834
            = *vernale* Pictet, 1834; synonym

conformis Neboiss, 1963
> = *boltoni*; misidentified by some earlier authors

intermedium (Klapálek, 1892)

**SYNAGAPETUS**
> dubitans McLachlan, 1879

**Superfamily Hydroptiloidea**
**Family HYDROPTILIDAE**

> **AGRAYLEA** Curtis, 1834
>> multipunctata Curtis, 1834
>> sexmaculata Curtis, 1834
>>> = *pallidula* McLachlan, 1875; synonym

> **ALLOTRICHIA** McLachlan, 1880
>> pallicornis (Eaton, 1873)

> **HYDROPTILA** Dalman, 1819
>> angulata Mosely, 1922
>> cornuta Mosely, 1922
>> forcipata (Eaton, 1873)
>> lotensis Mosely, 1930
>> martini Marshall, 1977
>> occulta (Eaton, 1873)
>> pulchricornis Pictet, 1834
>> simulans Mosely, 1920
>> sparsa Curtis, 1834
>> sylvestris Morton, 1898
>> tigurina Ris, 1894
>> tineoides Dalman, 1819
>>> = *femoralis* (Eaton, 1873); synonym
>>> = *longispina* McLachlan, 1884; synonym
>> valesiaca Schmid, 1947
>> vectis Curtis, 1834
>>> = *maclachlani* Klapálek, 1890; synonym

> **ITHYTRICHIA** Eaton, 1873
>> clavata Morton, 1905
>> lamellaris Eaton, 1873

> **ORTHOTRICHIA** Eaton, 1873
>> angustella (McLachlan, 1865)
>> costalis (Curtis, 1834)
>>> = *tetensii* Kolbe, 1887; synonym
>> tragetti Mosely, 1930

> **OXYETHIRA** Eaton, 1873
>> distinctella McLachlan, 1880
>> falcata Morton, 1893
>> flavicornis (Pictet, 1834)
>>> = *costalis* Eaton; synonym
>> frici Klapálek, 1891
>> mirabilis Morton, 1904
>> sagittifera Ris, 1897
>> simplex Ris, 1897
>> tristella Klapálek, 1895

**TRICHOLEIOCHITON** Kloet & Hincks, 1944
fagesii (Guinard, 1879)

**SUBORDER ANNULIPALPIA**
**Superfamily Philopotamoidea**
**Family PHILOPOTAMIDAE**
**CHIMARRA** Stephens, 1829
marginata (Linnaeus, 1761)
**PHILOPOTAMUS** Stephens, 1829
montanus (Donovan, 1813)
**WORMALDIA** McLachlan, 1865
mediana McLachlan, 1878
occipitalis (Pictet, 1834)
subnigra McLachlan, 1865

**Superfamily Hydropsychoidea**
**Family ECNOMIDAE**
**ECNOMUS** McLachlan, 1864
tenellus (Rambur, 1842)

**Family POLYCENTROPODIDAE**
**CYRNUS** Stephens, 1836
flavidus McLachlan, 1864
insolutus McLachlan, 1878
trimaculatus (Curtis, 1834)
**HOLOCENTROPUS** McLachlan, 1878
dubius (Rambur, 1842)
picicornis (Stephens, 1836)
stagnalis (Albarda, 1874)
**NEURECLIPSIS** McLachlan, 1864
bimaculata (Linnaeus, 1758)
**PLECTROCNEMIA** Stephens, 1836
brevis McLachlan, 1871
conspersa (Curtis, 1834)
geniculata McLachlan, 1871
**POLYCENTROPUS** Curtis, 1835
flavomaculatus (Pictet, 1834)
= *multiguttatus* Curtis, 1835; synonym
irroratus (Curtis, 1835)
= *multiguttatus*; misidentified by some earlier authors
kingi McLachlan, 1881

**Family PSYCHOMYIIDAE**
**LYPE** McLachlan, 1878
phaeopa (Stephens, 1836)
reducta (Hagen, 1868)
**PSYCHOMYIA** Latreille, 1829
= *METALYPE* Klapálek, 1898; synonym
fragilis (Pictet, 1834)
pusilla (Fabricius, 1781)

**TINODES** Leach, 1815
assimilis McLachlan, 1865
dives (Pictet, 1834)
maclachlani Kimmins, 1966
= *pusillus* Walker, 1852; synonym
maculicornis (Pictet, 1834)
pallidulus McLachlan, 1878
rostocki McLachlan, 1878
unicolor (Pictet, 1834)
waeneri (Linnaeus, 1758)

**Family HYDROPSYCHIDAE**
**CHEUMATOPSYCHE** Wallengren, 1891
lepida (Pictet, 1834)
**DIPLECTRONA** Westwood, 1840
felix McLachlan, 1878
**HYDROPSYCHE** Pictet, 1834
angustipennis (Curtis, 1834)
bulgaromanorum Malicky, 1977
= *guttata*; misidentified by some earlier authors
contubernalis McLachlan, 1865
= *ornatula* McLachlan, 1878; synonym
exocellata Dufour, 1841
fulvipes (Curtis, 1834)
instabilis (Curtis, 1834)
pellucidula (Curtis, 1834)
saxonica McLachlan, 1884
siltalai Döhler, 1963

**SUBORDER INTEGRIPALPIA**
**Superfamily Phryganeoidea**
**Family PHRYGANEIDAE**
**AGRYPNETES** McLachlan, 1876
crassicornis (McLachlan, 1876)
**AGRYPNIA** Curtis, 1835
obsoleta (Hagen, 1864)
pagetana Curtis, 1835
picta Kolenati, 1848
varia (Fabricius, 1793)
**HAGENELLA** Martynov, 1924
clathrata (Kolenati, 1848)
**OLIGOTRICHA** Rambur, 1842
striata (Linnaeus, 1758)
= *ruficrus* (Scopoli, 1763); synonym
**PHRYGANEA** Linnaeus, 1758
bipunctata Retzius, 1783
= *striata;* misidentified by some earlier authors
grandis Linnaeus, 1758
**TRICHOSTEGIA** Kolenati, 1848
minor (Curtis, 1834)

**Superfamily Limnephiloidea**
**Family BRACHYCENTRIDAE**
        **BRACHYCENTRUS** Curtis, 1834
           subnubilus Curtis, 1834

**Family GOERIDAE**
        **GOERA** Stephens, 1829
           pilosa (Fabricius, 1775)
        **SILO** Curtis, 1833
           nigricornis (Pictet, 1834)
           pallipes (Fabricius, 1781)

**Family LEPIDOSTOMATIDAE**
        **CRUNOECIA** McLachlan, 1876
           irrorata (Curtis, 1834)
        **LEPIDOSTOMA** Rambur, 1842
             = *LASIOCEPHALA* Costa, 1857; synonym
           basale (Kolenati, 1848)
           hirtum (Fabricius, 1775)
             = *fimbriatum*; misidentified by some earlier authors

**Family APATANIIDAE**
        **APATANIA** Kolenati, 1848
           auricula (Forsslund, 1930)
           muliebris McLachlan, 1866
             = *nielseni*; misidentified by some British authors
           wallengreni McLachlan, 1871

**Family LIMNEPHILIDAE**
  **Subfamily DICOSMOECINAE**
        **IRONOQUIA** Banks, 1916
           dubia (Stephens, 1837)
  **Subfamily DRUSINAE**
        **DRUSUS** Stephens, 1837
           annulatus (Stephens, 1837)
        **ECCLISOPTERYX** Kolenati, 1848
           dalecarlica Kolenati 1848
  **Subfamily LIMNEPHILINAE**
    **Tribe Chaetopterygini**
        **CHAETOPTERYX** Stephens, 1829
           villosa (Fabricius, 1798)
    **Tribe Limnephilini**
        **ANABOLIA** Stephens, 1837
             = *PHACOPTERYX* Kolenati, 1848; synonym
           brevipennis (Curtis, 1834)
           nervosa (Curtis, 1834)
        **GLYPHOTAELIUS** Stephens, 1837
           pellucidus (Retzius, 1783)

**GRAMMOTAULIUS** Kolenati, 1848
    nigropunctatus (Retzius, 1783)
       = *atomarius* (Fabricius, 1793); synonym
    nitidus (Müller, 1764)
**LIMNEPHILUS** Leach, 1815
    affinis Curtis, 1834
    auricula Curtis, 1834
    binotatus Curtis, 1834
       = *xanthodes* McLachlan, 1875; synonym
    bipunctatus Curtis, 1834
    borealis (Zetterstedt, 1840)
    centralis Curtis, 1834
    coenosus Curtis, 1834
    decipiens (Kolenati, 1848)
    elegans Curtis, 1834
    extricatus McLachlan, 1865
    flavicornis (Fabricius, 1787)
    fuscicornis (Rambur, 1842)
    fuscinervis (Zetterstedt, 1840)
    griseus (Linnaeus, 1758)
    hirsutus (Pictet, 1834)
    ignavus McLachlan, 1865
    incisus Curtis, 1834
    lunatus Curtis, 1834
    luridus Curtis, 1834
    marmoratus Curtis, 1834
    nigriceps (Zetterstedt, 1840)
    pati O'Connor, 1980
    politus McLachlan, 1865
    rhombicus (Linnaeus, 1758)
    sparsus Curtis, 1834
    stigma Curtis, 1834
    subcentralis (Brauer, 1857)
    tauricus Schmid, 1964
    vittatus (Fabricius, 1798)
**NEMOTAULIUS** Banks, 1906
    punctatolineatus (Retzius, 1783)
**RHADICOLEPTUS** Wallengren, 1891
    alpestris (Kolenati, 1848)
**Tribe Stenophylacini**
    **ALLOGAMUS** Schmid, 1955
        auricollis (Pictet, 1834)
    **ENOICYLA** Rambur, 1842
        pusilla (Burmeister, 1839)
    **HALESUS** Stephens, 1836
        digitatus (Schrank, 1781)
        radiatus (Curtis, 1834)
    **HYDATOPHYLAX** Wallengren, 1891
        infumatus (McLachlan, 1865)

**MELAMPOPHYLAX** Schmid, 1955
mucoreus (Hagen, 1861)
= *guttatipennis* McLachlan, 1865; synonym
**MESOPHYLAX** McLachlan, 1882
aspersus (Rambur, 1842)
impunctatus zetlandicus McLachlan, 1884
**MICROPTERNA** Stein, 1874
lateralis (Stephens, 1837)
sequax McLachlan, 1875
**POTAMOPHYLAX** Wallengren, 1891
cingulatus (Stephens, 1837)
= *latipennis*; misidentified by some earlier authors
latipennis (Curtis, 1834)
= *stellatus* (Curtis, 1834); synonym
rotundipennis (Brauer, 1857)
**STENOPHYLAX** Kolenati, 1848
permistus McLachlan, 1895
= *concentricus* McLachlan, 1875; synonym
vibex (Curtis, 1834)

**Superfamily Sericostomatoidea**
**Family SERICOSTOMATIDAE**
**NOTIDOBIA** Stephens, 1829
ciliaris (Linnaeus, 1761)
**SERICOSTOMA** Latreille, 1825
personatum (Spence, 1826)

**Family BERAEIDAE**
**BERAEA** Stephens, 1833
maurus (Curtis, 1834)
pullata (Curtis, 1834)
**BERAEODES** Eaton, 1867
minutus (Linnaeus, 1761)
**ERNODES** Wallengren, 1891
articularis (Pictet, 1834)

**Superfamily Leptoceroidea**
**Family ODONTOCERIDAE**
**ODONTOCERUM** Leach, 1815
albicorne (Scopoli, 1763)

**Family MOLANNIDAE**
**MOLANNA** Curtis, 1834
albicans (Zetterstedt, 1840)
= *palpata* McLachlan, 1877; synonym
angustata Curtis, 1834

**Family LEPTOCERIDAE**
**ADICELLA** McLachlan, 1877
filicornis (Pictet, 1834)
reducta (McLachlan, 1865)

**ATHRIPSODES** Billberg, 1820
    albifrons (Linnaeus, 1758)
    aterrimus (Stephens, 1836)
    bilineatus (Linnaeus, 1758)
    cinereus (Curtis, 1834)
    commutatus (Rostock, 1874)
**CERACLEA** Stephens, 1829
    albimacula (Rambur, 1842)
        = *alboguttatus* (Hagen, 1860); synonym
    annulicornis (Stephens, 1836)
    dissimilis (Stephens, 1836)
    fulva (Rambur, 1842)
    nigronervosa (Retzius, 1783)
    senilis (Burmeister, 1839)
**EROTESIS** McLachlan, 1877
    baltica McLachlan, 1877
**LEPTOCERUS** Leach, 1815
    interruptus (Fabricius, 1775)
    lusitanicus (McLachlan, 1884)
    tineiformis Curtis, 1834
**MYSTACIDES** Latreille, 1825
    azurea (Linnaeus, 1761)
    longicornis (Linnaeus, 1758)
    nigra (Linnaeus, 1758)
**OECETIS** McLachlan, 1877
    furva (Rambur, 1842)
    lacustris (Pictet, 1834)
    notata (Rambur, 1842)
    ochracea (Curtis, 1825)
    testacea (Curtis, 1834)
**SETODES** Rambur, 1842
    argentipunctellus McLachlan, 1877
    punctatus (Fabricius, 1793)
**TRIAENODES** McLachlan, 1865
    bicolor (Curtis, 1834)
**YLODES** Milne, 1934
    conspersus (Rambur, 1842)
    reuteri (McLachlan, 1880)
        = *simulans*; misidentified by some earlier authors
    simulans (Tjeder, 1929)

# Key to families

It is recommended that the newcomer to the Trichoptera should read the introductory section in this handbook before attempting to use this key to the 19 families. All the relevant characters are explained and discussed there, and the relative merits of using the key or simply comparing illustrations are also outlined.

Family keys have traditionally separated the tiny Hydroptilidae from the other families by the presence of long fringes of hairs on the hind wings; however, some members of the Beraeidae and smaller members of the Leptoceridae also show this feature. Therefore these two families key out in two different places. Note that the wingless female of *Enoicyla pusilla* will not key out here, but can hardly be mistaken for anything else (p. 148).

In all keys in this handbook a number in brackets after a couplet number indicates how this couplet was reached, e.g. 12(4) means that couplet 4 led to couplet 12. This can be useful for back-tracking.

1      Some hairs fringing hind wing are longer than greatest breadth of wing (Fig. 15) ................................................. 2

15

-      All hairs fringing hind wing are shorter than greatest breadth of wing (Fig. 16) ............................................... 4

2(1) Only 2 spurs on hind leg ................ **Leptoceridae** (p. 158)

-      4 spurs on hind leg .......................................................... 3

16

3(2) Two spurs on fore leg; row of spines present on mid and hind legs; fore wings rounded (Fig. 16) ........ **Beraeidae** (p. 152)

-      No spurs on fore leg; no row of spines on legs; fore wings pointed (Fig. 15) ............................ **Hydroptilidae** (p. 36)

4(1) Mid and hind tibiae with dark spines that are more than half as long as paler spurs (Fig. 17) ........................................ 5

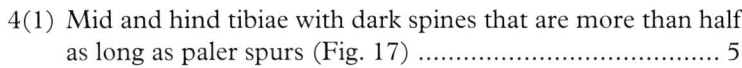

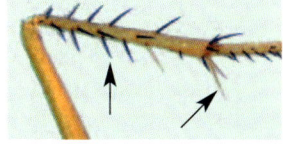

17

-      Mid and hind tibiae with spines that are less than half length of spurs (Fig. 18) or else spines absent ............................. 7

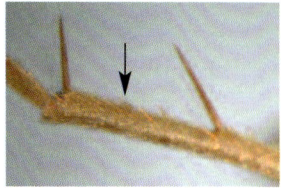

18

5(4) Spur formula 2.4.4 or 1.2.2; basal antennal segment about twice length of next segment; basal antennal segment only slightly longer than broad (Fig. 19) ..... **Phryganeidae** (p. 90)

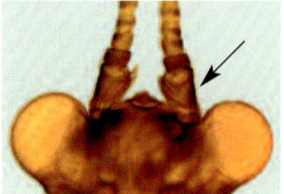

19

- Spur formula never 2.4.4 or 1.2.2; basal antennal segment about three times length of next segment; basal antennal segment about 1.5-3 times longer than broad (Fig. 20) ..... 6

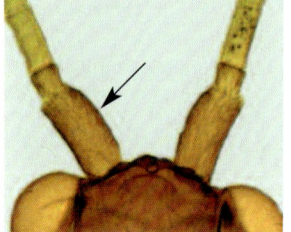

20

6(5) Pterostigma in fore wing forming bulge on front margin. In hind wing discoidal cell open (Fig 21). Spur formula 1.2.4 ........................................... **Apataniidae** (p. 110)

- Pterostigma not forming bulge on front margin, even if large and strongly marked. In hind wing discoidal cell closed (Fig. 22). Spur formula varies but never 1.2.4. .............. .................................................................................. **Limnephilidae** (p. 112)

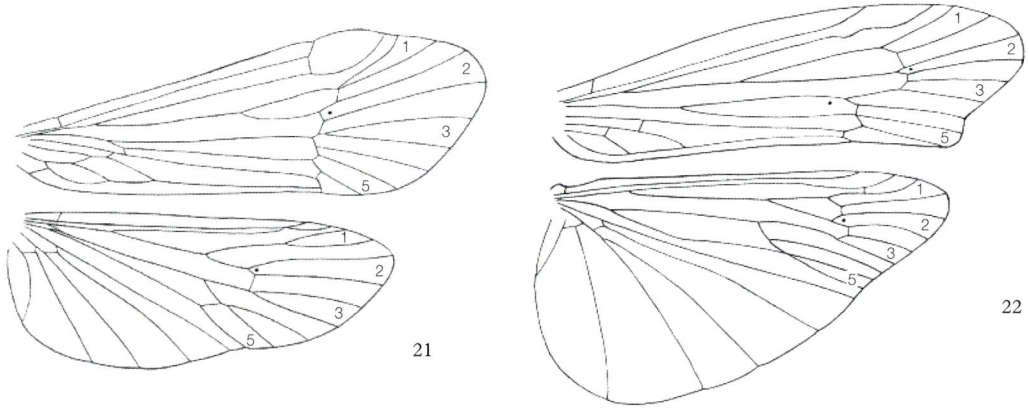

21

22

7(4) Spur formula 3.4.4 ..................................................................................... 8

- Spur formula not 3.4.4 ............................................................................. 10

8(7) Ocelli present ................................................................. **Rhyacophilidae** (p. 27)

- Ocelli absent ............................................................................................. 9

9(8)  Vein R1 in fore wing not forked (Fig. 23); pair of fine lateral filaments present on 5th abdominal segment ........................................................ **Polycentropodidae** (p. 61)

-     Vein R1 in fore wing forked (Fig. 24); no lateral filaments on abdomen ........................ ................................................................................................ **Ecnomidae** (p. 60)

23

24

10(7)  Spur formula 1.4.4 or 2.4.4 ............................................................................... 11

-     Spur formula not 1.4.4 or 2.4.4 ....................................................................... 18

11(10) Basal antennal segment at least 4 times length of next segment (Fig. 25) ...................................................... 12

25

-     Basal antennal segment less than 4 times length of next segment (Fig. 26) ...................................................... 14

26

12(11) Hind margin of hind wing rather straight and parallel to costa, giving wing a subrectangular appearance (Fig. 27) ...................................................... **Goeridae** (p. 104)

27

-     Hind margin of hind wing evenly curved (Fig. 28) ...... 13

28

13(12) Cross-vein present between veins R1 and R2 in fore wing (Fig. 29); fore wing length more than 11 mm .......................................................... **Odontoceridae** (p. 156)

- No cross-vein between R1 and R2 in fore wing (Fig. 30); fore wing length less than 10 mm ................................................................. **Lepidostomatidae** (p. 107)

29

30

14(11) In fore wing discoidal cell open (Fig. 31) ............................. **Molannidae** (p. 157)

- In fore wing discoidal cell closed (Fig. 32) ......................................................... 15

31

32

15(14) In fore wing median cell open (Fig. 33) .......................... **Glossosomatidae** (p. 30)

- In fore wing median cell closed (Fig. 34) ........................................................... 16

33

34

16(15) Ocelli present ................................................................. **Philopotamidae** (p. 54)

- Ocelli absent ................................................................................................ 17

17(16) In fore wing Fork 1 absent (Fig. 35) ................................. **Psychomyiidae** (p. 72)

- In fore wing Fork 1 present (Fig. 36) ............................... **Hydropsychidae** (p. 81)

35

36

18(10)  Only two spurs on hind tibia ............................................. **Leptoceridae** (p. 158)

-  More than two spurs on hind tibia ..................................................................... 19

19(18)  Spur formula 2.3.3 ..................................................... **Brachycentridae** (p. 102)

-  Spur formula 2.2.4 ......................................................................................... 20

20 (19)  Fore wing length 6 mm or less ............................................... **Beraeidae** (p. 152)

-  Fore wing length 9 mm or more ................................... **Sericostomatidae** (p. 150)

# Family RHYACOPHILIDAE (1 genus, 4 species)

This family includes the single genus *Rhyacophila*, with just four species in Britain, though there are around 90 species throughout Europe. *R. dorsalis* is by far the most common and widespread, but the other three can be locally common in suitable areas; all live in stony rivers and streams. All species have a similar wing venation (Fig. 37) with forks 1-5 present in the fore wing, and 1, 2, 3 and 5 in the hind wing; discoidal cell open in both wings. There is no sexual dimorphism in colour or venation although minor venational variations are occasionally seen. Spur formula 3.4.4 in both sexes. Ocelli present.

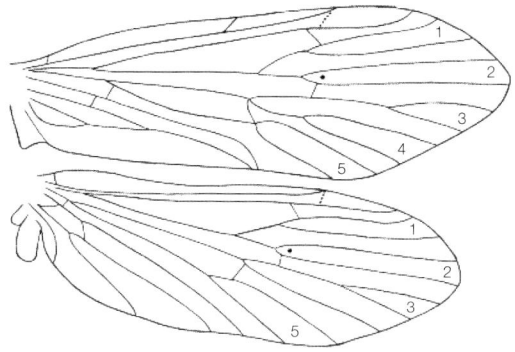

Figure 37. *Rhyacophila dorsalis.* Wing venation

## Genus RHYACOPHILA Pictet, 1834

### *Rhyacophila dorsalis* (Curtis, 1834)

This is known to anglers as the Sand Fly. Fore wing mottled grey and yellow, but very variable in both colour and pattern (Fig. 38); living specimens show a distinct dark diamond shape where the wings meet over the body at rest. Fore wing-length: ♂ 10-12 mm, ♀ 11-14 mm. Common throughout most of Britain, present in Ireland; stony flowing water ranging from small streams to large rivers. According to Malicky (2002) the subspecies *R. dorsalis dorsalis* occurs only in the British Isles and some parts of western Europe, with other subspecies occurring elsewhere in Europe. Flight period: May-October, but throughout the whole year in southern chalk rivers; probably bivoltine. ♂ genitalia with apical segment of clasper almost rectangular in lateral view (Fig. 39); ♀ genitalia with narrow V-shaped ventral excision in segment VIII, internal plate large and triangular (Fig. 40).

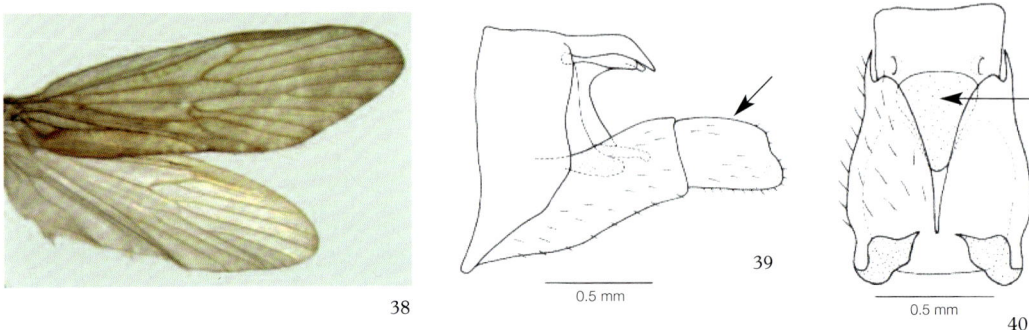

Figures 38-40. *Rhyacophila dorsalis*. 38 wing pattern; 39 male genitalia lateral, aedeagus omitted; 40 female genitalia ventral, terminal segments omitted

## *Rhyacophila fasciata* Hagen 1859
= *septentrionis* McLachlan, 1865

*R. fasciata* is a variable species, possibly including several subspecies throughout Europe, and its status still needs further work. However, *septentrionis* is currently regarded as a synonym (Malicky, 2005) so, for the time being, we must accept *fasciata* as the valid name for the British representatives. Fore wing greyish yellow with variable pattern (Fig. 41). Fore wing length: ♂♀ 10-14 mm. Scattered throughout Britain in streams that form calcareous deposits where it may be common, but also in smaller numbers in other stony alkaline streams and rivers; no records from Ireland. Throughout Europe, but may include several subspecies. Often day-flying, apparently not attracted to light. Flight period: June-September. ♂ genitalia with apical segment of clasper triangular in lateral view (Fig. 42); ♀ genitalia with two halves of segment VIII widely separated in ventral view, internal plates very small and paired (Fig. 43).

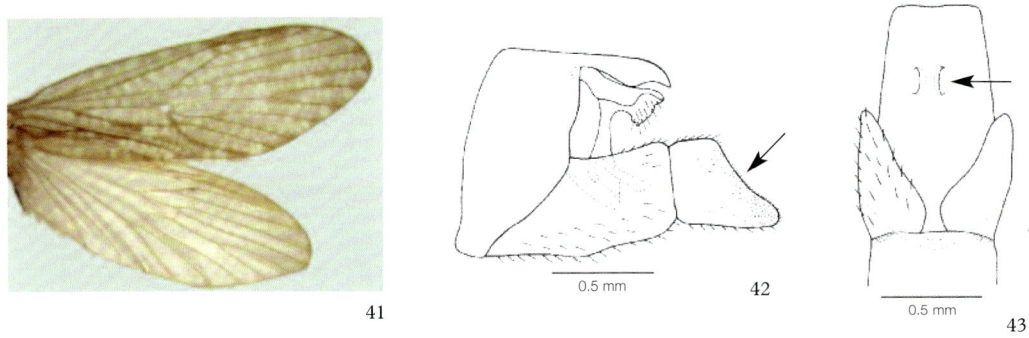

Figures 41-43. *Rhyacophila fasciata*. 41 wing pattern; 42 male genitalia lateral, aedeagus omitted; 43 female genitalia ventral, terminal segments omitted

## *Rhyacophila munda* McLachlan, 1862

Fore wing mottled grey and yellow in life, very similar to *R. dorsalis*, but easily separated by features of the genitalia in both sexes. Pterostigma in fore wing often dark (Fig. 44). Fore wing length: ♂ ♀ 10-13 mm. SW. and N. England, Wales and bordering counties of England, mainland Scotland; present in Ireland. Apparently not common and larvae are more often seen than adults; upland stony streams and rivers. Mainly south-western Europe. Flight period: (May) July-October (November); the late flight period may partly explain the paucity of adult records. ♂ genitalia with apical segment of clasper elongate and subdivided at tip (Fig. 45); ♀ genitalia with broad U-shaped ventral excision in segment VIII, internal plate small and triangular or subrectangular (Fig. 46).

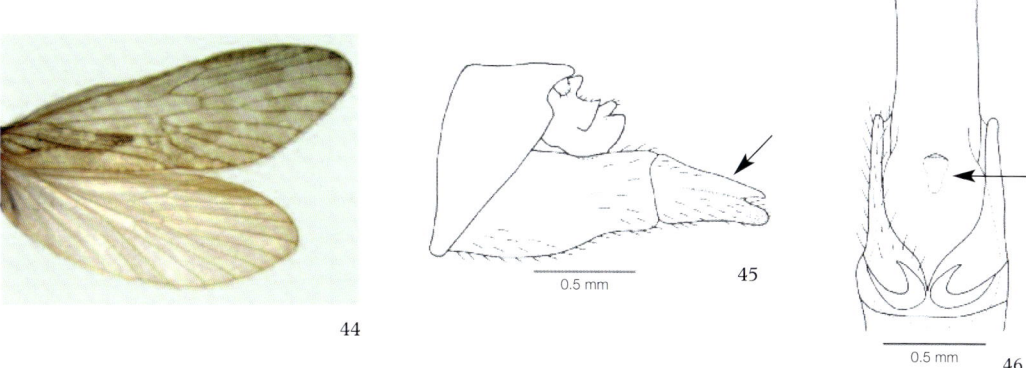

Figures 44-46. *Rhyacophila munda.* 44 wing pattern; 45 male genitalia lateral, aedeagus omitted;
46 female genitalia ventral, terminal segments omitted

## *Rhyacophila obliterata* McLachlan, 1863

Fore wing bright yellow in life, fading slightly after death, with slight grey pattern (Fig. 47). Fore wing length: ♂ ♀ 11-14 mm. Widespread in SW. and N. England, north Midlands, Wales and mainland Scotland, but usually not common; no records from Ireland; fast rivers and streams. Throughout much of Europe. Flight period: July-October; as with *R. munda* the late flight period may disguise local abundance at some sites. ♂ genitalia with ventrally elongated apical segment of clasper (Fig. 48); ♀ genitalia with broad U-shaped ventral excision in segment VIII, internal plates small, elongate and paired (Fig. 49).

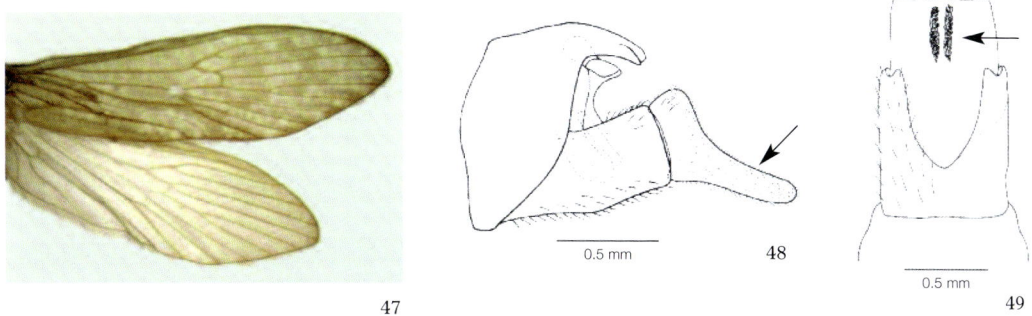

Figures 47-49. *Rhyacophila obliterata.* 47 wing pattern; 48 male genitalia lateral, aedeagus omitted;
49 female genitalia ventral, terminal segments omitted

# Family GLOSSOSOMATIDAE (3 genera, 7 species)

Fore wing with forks 1 to 5 present, hind wing with forks 1, 2, 3 and 5 (only 2, 3 and 5 in *Agapetus* and *Synagapetus*); discoidal cell closed in fore wing, open or closed in hind wing. Spur formula 2.4.4. Ocelli present. The tibiae and tarsi of the middle legs in the female are broad and flattened.

## Key to the genera of Glossosomatidae

1. In hind wing fork 1 present; discoidal cell present (Fig. 50, dc); fore wing usually at least 7 mm long ................................................................................. *Glossosoma* (p. 30)

- In hind wing fork 1 absent; discoidal cell absent (Fig. 51); fore wing usually no more than 6 mm long .................................................................................................... 2

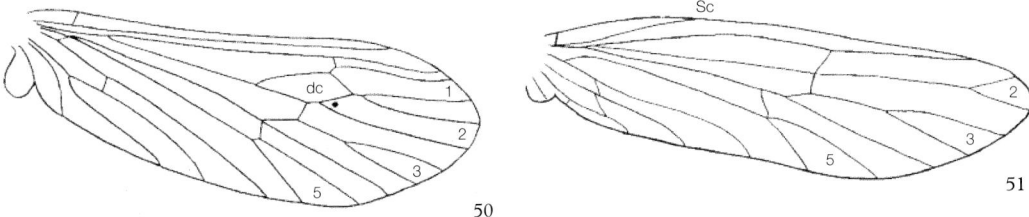

2. In hind wing subcosta very short, less than third of wing length (Fig. 51, Sc) ...................
............................................................................................................ *Agapetus* (p. 33)

- In hind wing subcosta more than half length of wing (Fig. 52, Sc) ...... *Synagapetus* (p. 35)

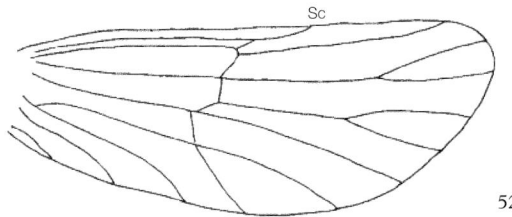

# Genus GLOSSOSOMA Curtis, 1834

The three British species in this genus are all associated with fast-running water, often in upland habitats. There are about 25 species in Europe. Discoidal cell closed in both wings (Fig. 53). All species show sexual dimorphism in that the males have a scent-scale structure at the base of the fore wing, visible to the naked eye. The nomenclature of this group has changed considerably since Mosely's (1939) book and was clarified by the re-examination of the Curtis collection by Neboiss (1963).

Note that the terminal segments of the female genitalia, which form an ovipositor, may be more or less evaginated; the critical segment to examine is the 8th (shaded in Figs 58, 63 and 67).

## *Glossosoma boltoni* **Curtis, 1834**

As a result of earlier confusion about the correct naming of some species in *Glossosoma*, this species was listed as *G. vernale* in Mosely (1939) but was correctly named in Macan (1973). Fore wing greyish yellow or brown with some indistinct paler markings (Figs 53-56); male with dark pouch at base of fore wing (Fig. 54). Fore wing length: ♂♀ 7-9 mm. Common in Scotland, Wales, N. and SW. England, present in Ireland; stony streams and rivers. Throughout most of Europe. Flight period: April-November, probably bivoltine. ♂ genitalia with a large telescopic left paramere, segment X with upturned tip (Fig. 57); ♀ genitalia with unmodified segment VIII (Fig. 58).

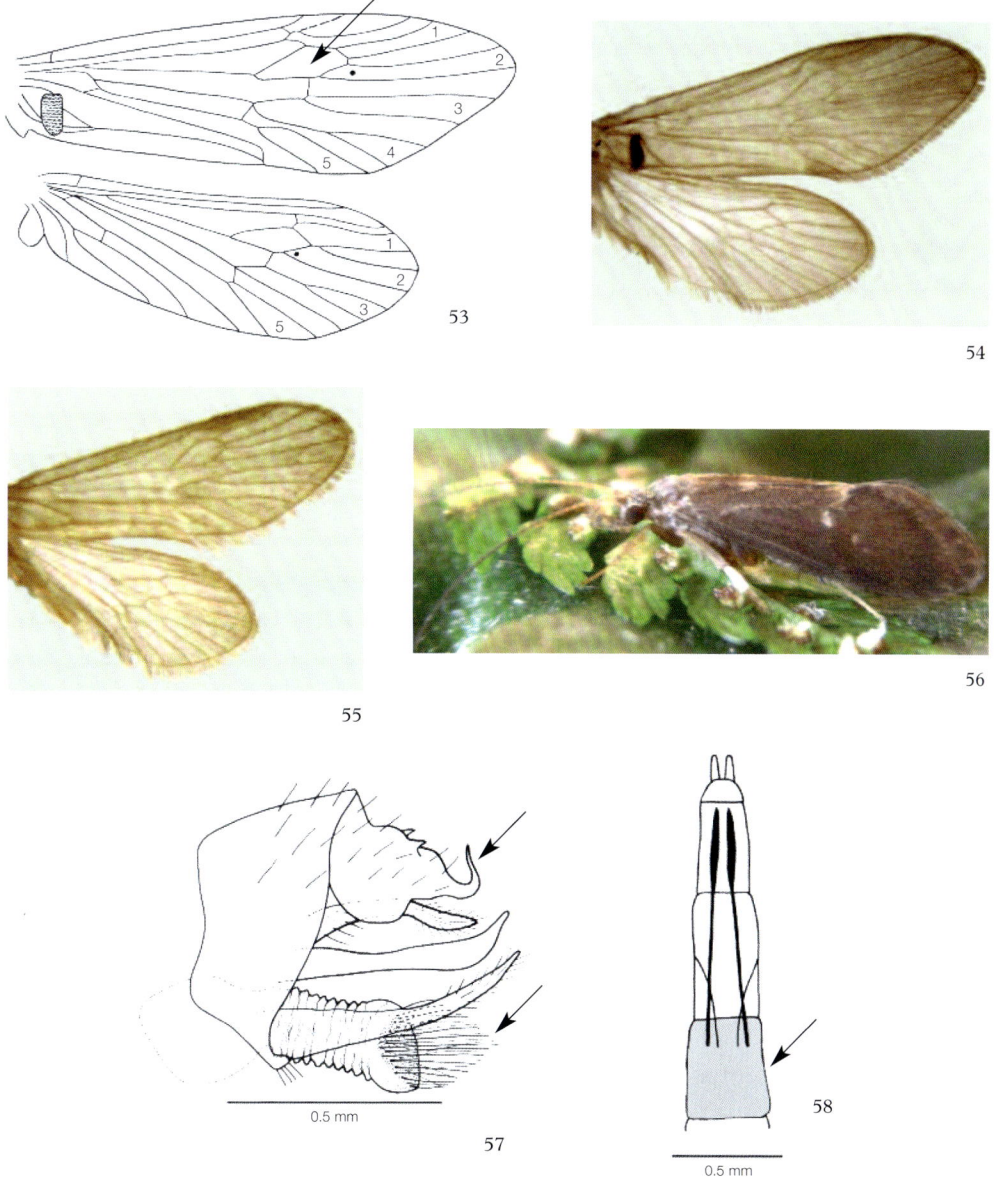

Figures 53-58. *Glossosoma boltoni.* 53 male wing venation; 54 male wing pattern; 55 female wing pattern; 56 live specimen; 57 male genitalia lateral; 58 female genitalia ventral, segment VIII shaded

31

## *Glossosoma conformis* Neboiss, 1963

This species was listed as *G. boltoni* in Mosely (1939) but was correctly named in Macan (1973). Fore wing similar to *G. boltoni*. (Figs 59-61). Fore wing length: ♂ 7-8 mm, ♀ 8-10 mm. Common in Scotland, Wales, N. and SW. England, present in Ireland; stony streams and rivers. Throughout much of Europe. Flight period: May-August. ♂ genitalia with a large left paramere like *G. boltoni* but segment X with downturned tip (Fig. 62); ♀ genitalia with ventral division on segment VIII and rounded margins (Fig. 63).

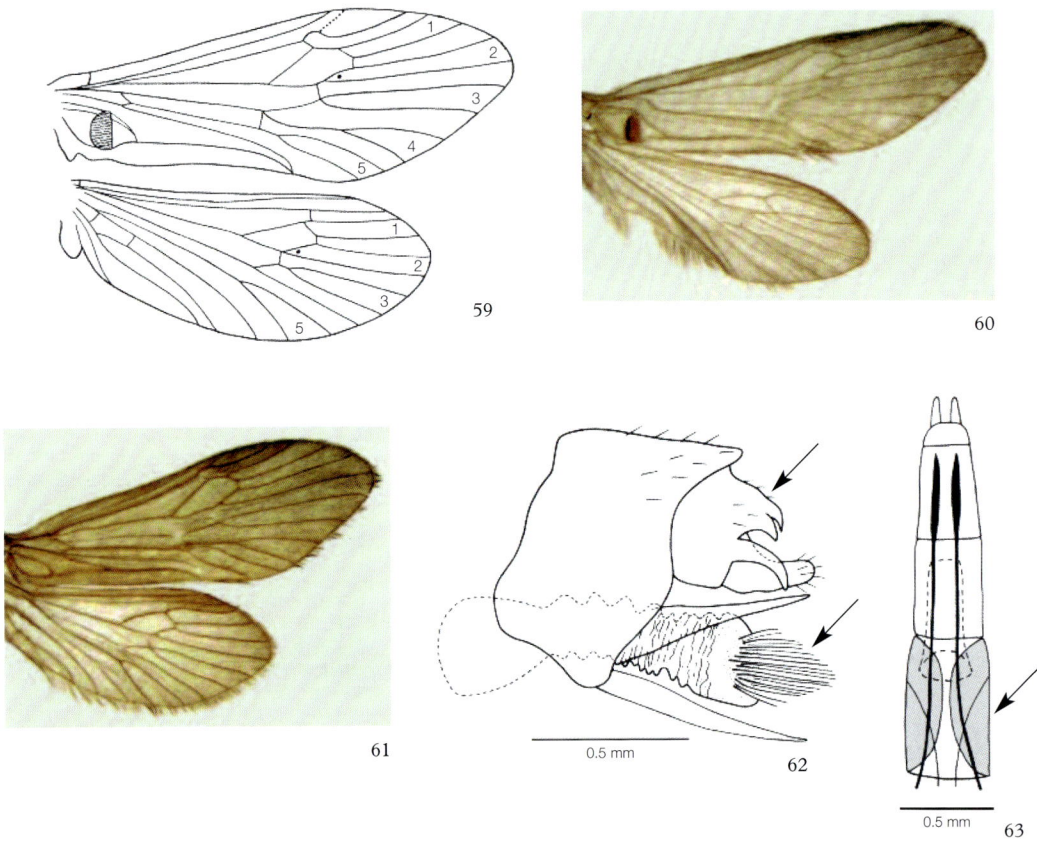

Figures 59-63. *Glossosoma conformis*. 59 male wing venation; 60 male wing pattern; 61 female wing pattern; 62 male genitalia lateral; 63 female genitalia ventral, segment VIII shaded

## *Glossosoma intermedium* (Klapálek, 1892)

Listed as *Mystrophora intermedia* in Mosely (1939). This is a BAP species: the Small Grey Sedge. Fore wing colour similar to the other species of *Glossosoma* but the scent-organ on the male wing is in the form of a brownish patch of dispersed scent-scales, rather than a distinct pouch (Figs 64, 65). Fore wing length: ♂ 7 mm, ♀ 8 mm. Lake District only; no records from Ireland; fast, stony streams. Central and northern Europe. Flight period: April-May. ♂ genitalia without enlarged parameres, segment X with long pointed dorsal processes (Fig.66); ♀ genitalia with ventral division on segment VIII and sinuate anterior and posterior margins (Fig. 67).

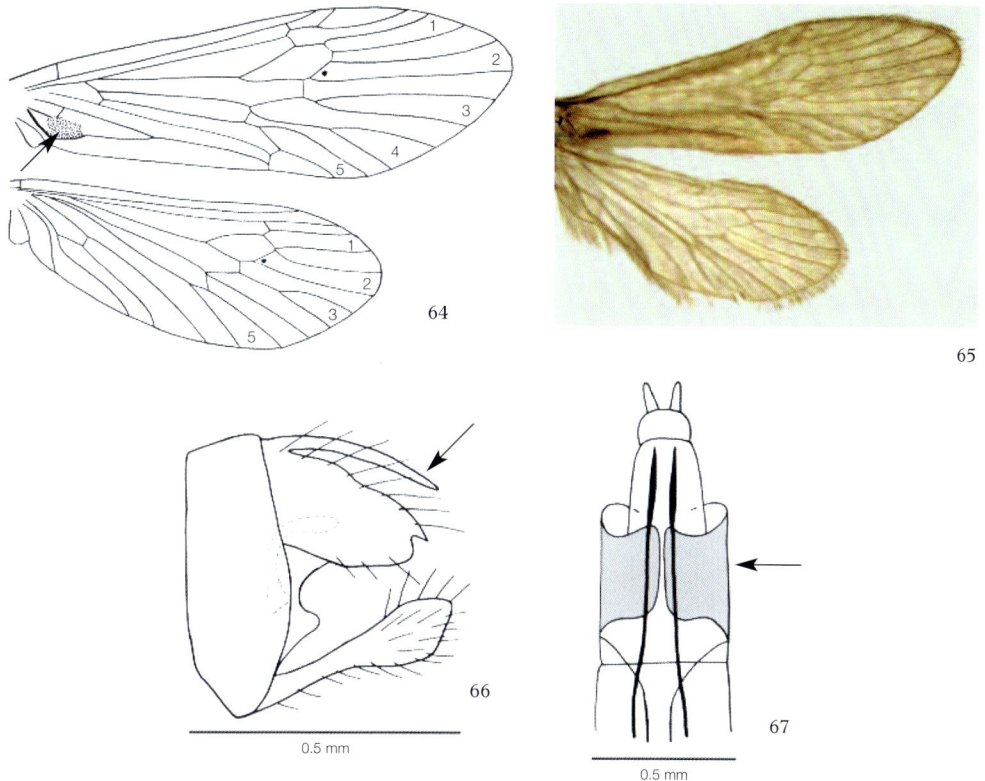

Figures 64-67. *Glossosoma intermedium*. 64 male wing venation; 65 male wing pattern;
66 male genitalia lateral; 67 female genitalia ventral, segment VIII shaded

# Genus AGAPETUS Curtis, 1834

The three British species in this genus are all associated with running water and often occur in huge numbers. These are the angler's Tiny Grey Sedges. There are over 30 species in Europe. Discoidal cell closed in fore wing, open in hind wing (Fig. 68). No sexual dimorphism in wing patterning or venation.

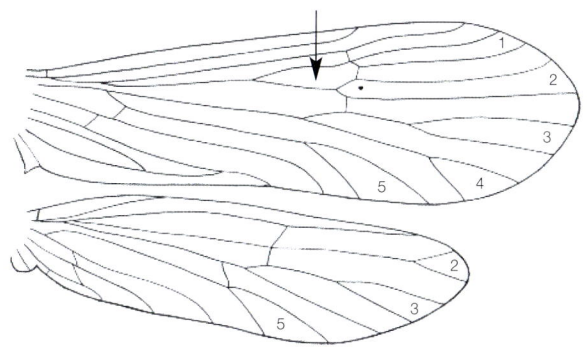

Figure 68. *Agapetus fuscipes*. Wing venation

## *Agapetus fuscipes* Curtis, 1834

Fore wing dark greyish brown with yellow hairs in life, but fading to a yellowish grey in dried museum specimens (Fig. 69). Fore wing length: ♂ 4-5 mm, ♀ 4-6 mm. Common throughout Britain, present in Ireland; stony bottomed flowing water, ranging from small trickles to large rivers though commonest in smaller water-bodies. Adults day-flying, not usually attracted to light. Flight period: April-October, probably bivoltine. Throughout much of Europe. ♂ genitalia with strongly upturned tip to segment X (Fig. 70) and claspers with teeth on inner margin (Fig. 71); ♀ genitalia with posterior margin of segment VIII unmodified in lateral view (Fig. 72).

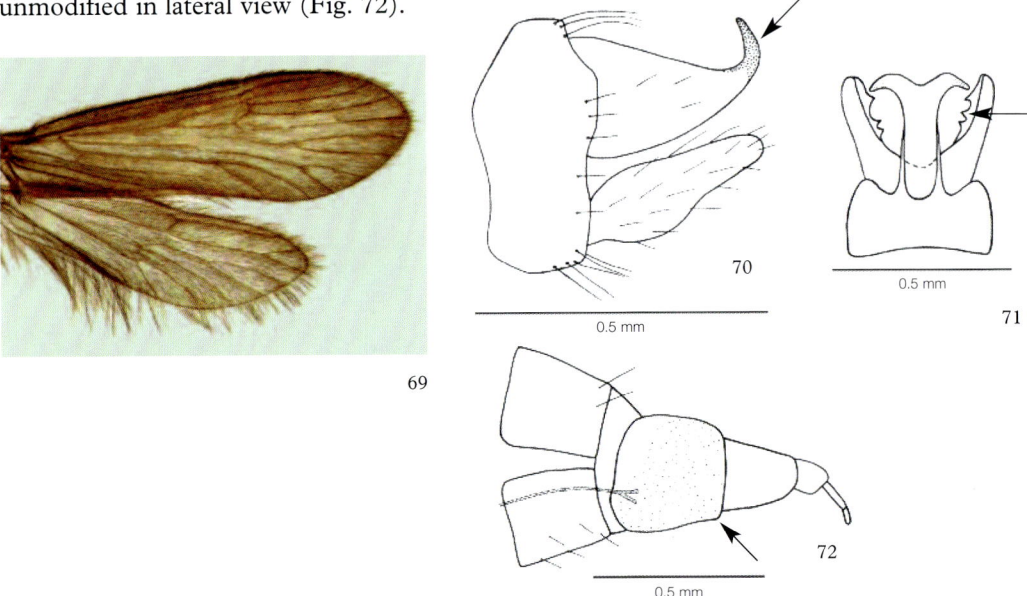

Figures 69-72. *Agapetus fuscipes.* 69 wing pattern; 70 male genitalia lateral, aedeagus omitted; 71 male genitalia dorsal; 72 female genitalia lateral, segment VIII stippled

## *Agapetus ochripes* Curtis, 1834

This was listed as *A. comatus* in Mosely (1939). Fore wing (Fig. 73) similar to *A. fuscipes* though paler in life. Fore wing length: ♂ 3-5 mm, ♀ 4-5 mm. Southern, central, western and northern England, Wales and mainland Scotland, present in Ireland; stony streams and rivers but commonest in larger water-bodies. Throughout Europe. Flight period: May-October, but bivoltine with two clear peaks in most areas (Cooling, 1982). ♂ genitalia with straight pointed tip to segment X, claspers with spines at apex (Fig. 74); ♀ genitalia with deeply incised posterior margin to segment VIII in lateral view (Fig. 75).

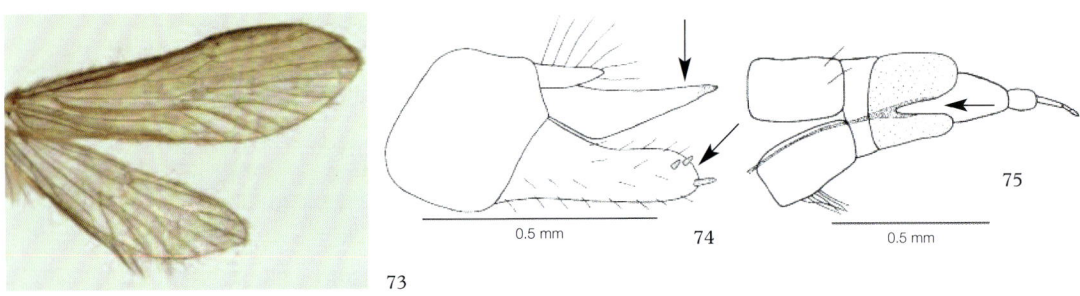

Figures 73-75. *Agapetus ochripes.* 73 wing pattern; 74 male genitalia lateral, aedeagus omitted; 75 female genitalia lateral, segment VIII stippled

## *Agapetus delicatulus* **McLachlan, 1884**

Fore wing similar to the other species of *Agapetus* (Fig. 76). Fore wing length: ♂ 4-5 mm, ♀ 5-6 mm. SW. and NW. England, Wales and bordering counties of England, mainland Scotland, locally common; present in Ireland; stony rivers and streams. Throughout Europe except extreme north. Flight period: May-August. ♂ genitalia with sinuate tip to segment X in lateral view (Fig. 77), and claspers toothed on inner margin though not as pronounced as in *A. fuscipes* (cf. Fig. 71); ♀ genitalia with shallow V-shaped incision on posterior margin of segment VIII in lateral view (Fig. 78).

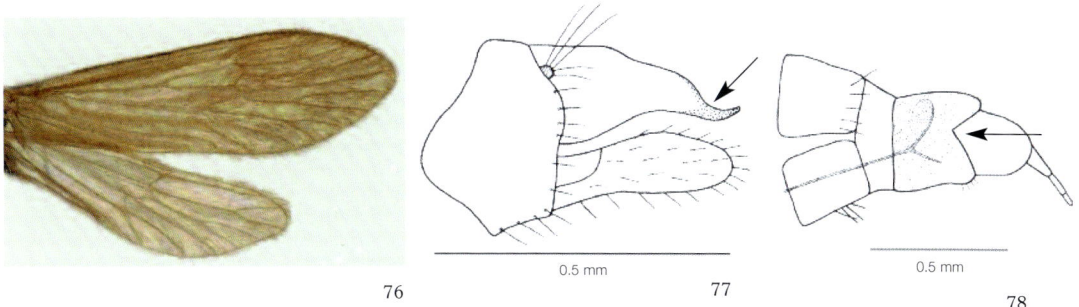

Figures 76-78. *Agapetus delicatulus.* 76 wing pattern;
77 male genitalia lateral, aedeagus omitted; 78 female genitalia lateral, segment VIII stippled

# Genus SYNAGAPETUS McLachlan, 1879

The single British species in this genus was discovered in 2010. There are several European species but most have a limited distribution. Discoidal cell closed in fore wing, open in hind wing, as in *Agapetus*, but with a much longer subcosta in the hind wing (Fig. 52). No sexual dimorphism in wing patterning or venation.

## *Synagapetus dubitans* **McLachlan, 1879**

This species is very similar to *Agapetus* in general size and appearance. It is currently known from only a single site in a small spring-fed stream in Yorkshire (Crofts, 2011). Its small size and limited dispersal abilities would suggest that it is an overlooked resident rather than a recent introduction. On the continent *S. dubitans* is found in calcareous streams in western and central Europe. ♂ genitalia with upturned tip and small dark spine near apex of segment X (Fig. 79); ♀ genitalia with rounded process on posterior margin of segment VIII (Fig. 80).

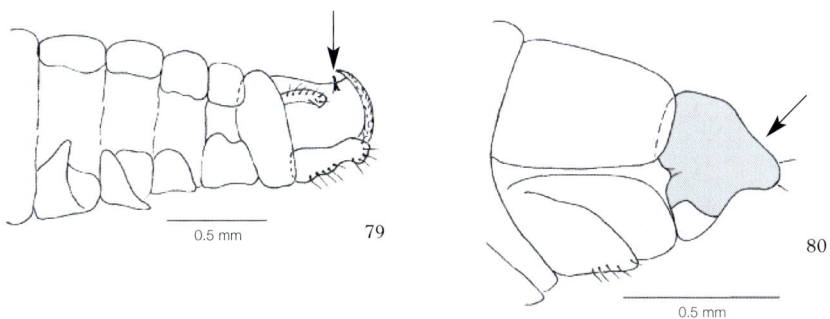

Figures 79-80. *Synagapetus dubitans.* 79 male genitalia lateral; 80 female genitalia lateral, segment VIII shaded

# Family HYDROPTILIDAE (7 genera, 31 species)

Spur formula 0.2.4 or 0.3.4; ocelli present or absent in different genera. These are the angler's Micro Caddis; because of their small size it is often necessary to slide-mount cleared specimens for examination under a compound microscope.

Genitalia drawings in this family are taken from Marshall (1978, 1979).

The following key to genera is not easy to use because of the difficulty in seeing the spur formula, ocelli and wing venation in these tiny insects. However, it can help to narrow down the identity of a specimen in those cases where browsing through the genitalia drawings of all 31 species in Britain seems an overwhelming prospect. Some pairs of smaller genera can only be separated by genitalic characters.

## Key to genera of Hydroptilidae

1. Spur formula 0.2.4 ..................................................................... *Hydroptila* (p. 37)

- Spur formula 0.3.4 ............................................................................................ 2

2. Ocelli absent ......................................................................... *Orthotrichia* (p. 47)

- Ocelli present ................................................................................................... 3

3. Fore wing length 4 mm or more .......................... *Agraylea* (p. 45) and *Allotrichia* (p. 46)

- Fore wing length 3.5 mm or less ..................................................................... 4

4. Both wings tapering narrowly to a sharp point (Fig. 81) ....................... *Oxyethira* (p. 50)

- Both wings gently tapering and rounded at apex (Fig. 82) ...........................................
................................................................ *Ithytrichia* (p. 44) and *Tricholeiochiton* (p. 49)

81

82

# Genus HYDROPTILA Dalman, 1819

A very large genus with almost 100 species in Europe, but just 14 in Britain. Spur formula 0.2.4. Ocelli absent.

## *Hydroptila tineoides* **Dalman, 1819**
= *femoralis* (Eaton, 1873)
= *longispina* McLachlan, 1884

This was listed as *H. femoralis* in Mosely (1939). Fore wing length: ♂ 2.5-3 mm, ♀ 3 mm. The male front femora are covered with black hairs. Common throughout Britain, present in Ireland; stony streams and lakes. Throughout Europe. Flight period: May-September. ♂ genitalia with dark downturned appendages on segment IX (Fig. 83); ♀ genitalia with two large setose ventral lobes (Fig. 84).

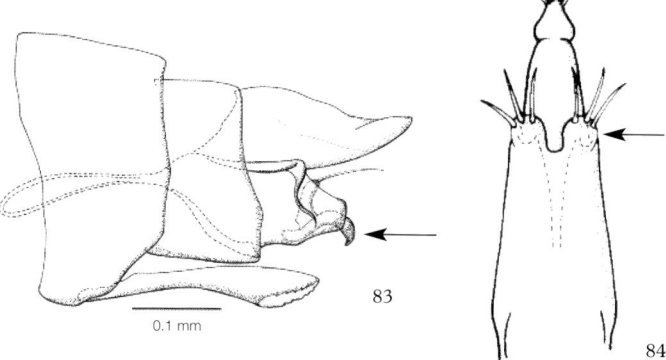

Figures 83-84. *Hydroptila tineoides.* 83 male genitalia lateral, aedeagus omitted; 84 female genitalia ventral

## *Hydroptila pulchricornis* **Pictet, 1834**

Fore wing length: ♂ 2.5 mm, ♀ 3 mm. Throughout Britain, not common; present in Ireland; lakes and rivers. Scattered throughout Europe except south. Flight period: May-September. ♂ genitalia very narrow in lateral view with long narrow claspers (Fig. 85), tergite X elongate and pointed in dorsal view (Fig. 86); ♀ genitalia with two setose lobes on dorsal side of segment VIII (Fig. 87), not ventral as in *H. tineoides*.

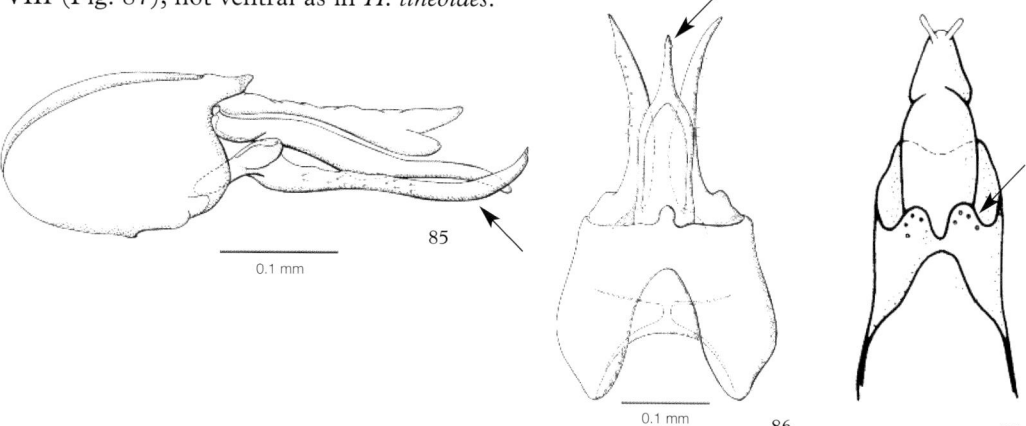

Figures 85-87. *Hydroptila pulchricornis.* 85 male genitalia lateral;
86 male genitalia dorsal, aedeagus omitted; 87 female genitalia dorsal

## *Hydroptila forcipata* (Eaton, 1873)

Fore wing length: ♂♀ 3-3.5 mm. Common throughout Britain; present in Ireland; stony streams and rivers. Throughout Europe. Flight period: April-September. ♂ genitalia with upcurved claspers and down-curved superior appendages, giving a forceps-like appearance in lateral view (Figs 88-89); ♀ genitalia with a pair of very narrow setose lobes on the ventral side of segment VIII (Fig. 90).

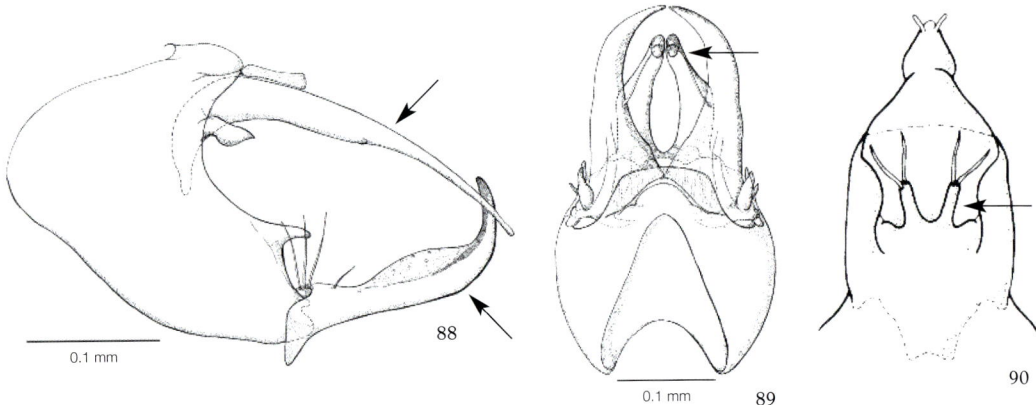

Figures 88-90. *Hydroptila forcipata.* 88 male genitalia lateral;
89 male genitalia dorsal, aedeagus omitted; 90 female genitalia ventral

## *Hydroptila vectis* Curtis, 1834
= *maclachlani* Klapálek, 1890

Listed as *H. maclachlani* in Mosely (1939). Fore wing length: ♂♀ 3-3.5 mm (Fig. 91). Common throughout Britain except N. Scotland, no records from Ireland; streams. Throughout Europe except extreme north. Flight period: June-September. ♂ genitalia with pair of black sinuous spines arising beneath segment X (Fig. 92); ♀ genitalia with ventral quadrate excision in posterior margin and small dark Y-shaped sclerite (Fig. 93).

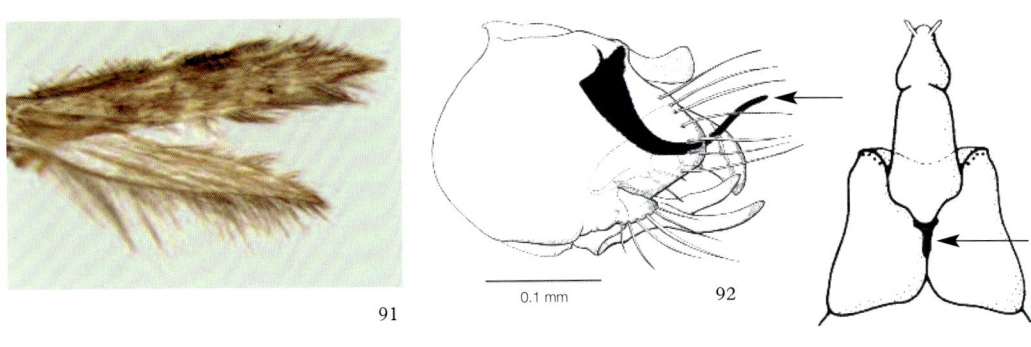

Figures 91-93. *Hydroptila vectis.* 91 wing pattern;
92 male genitalia lateral, aedeagus omitted; 93 female genitalia ventral

## *Hydroptila tigurina* **Ris, 1894**

Fore wing length: ♂♀ 3 mm. Lake District and northern Scotland, with a record from Ireland (O'Connor, 1978); fast rivers. Mainly southern Europe. Flight period: August. ♂ genitalia with toothed postero-lateral margins to segment IX (Fig. 94); ♀ genitalia asymmetrical, with ventral lobe on left side of segment VIII (Fig. 95). The female was originally known only from a single specimen, and it was not clear whether the genitalia were really asymmetrical or whether this specimen was aberrant (Marshall, 1979). Recently collected material from northern Scotland confirms that they are indeed asymmetrical (David Pryce, pers comm.).

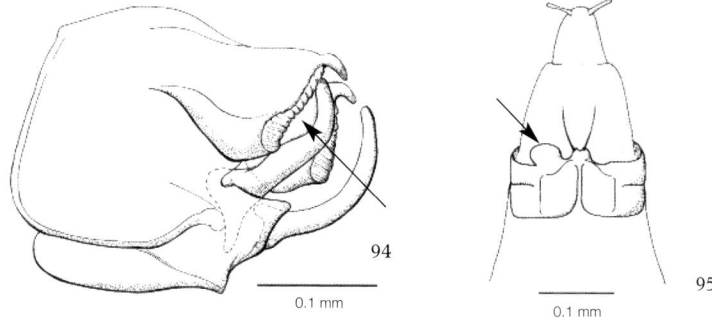

Figures 94-95. *Hydroptila tigurina*. 94 male genitalia lateral, aedeagus omitted; 95 female genitalia ventral

## *Hydroptila sylvestris* **Morton, 1898**

Fore wing length: ♂♀ 3-3.5 mm. Scotland and Lake District, not common; no records from Ireland; habitat not known. South-west Europe only. Flight period: June-August. ♂ genitalia with large triangular lateral processes on segment XI (Fig. 96), segment X rounded with small apical excision in dorsal view (Fig. 97); ♀ genitalia with deep and narrow dorsal excision on segment VIII (Fig. 98).

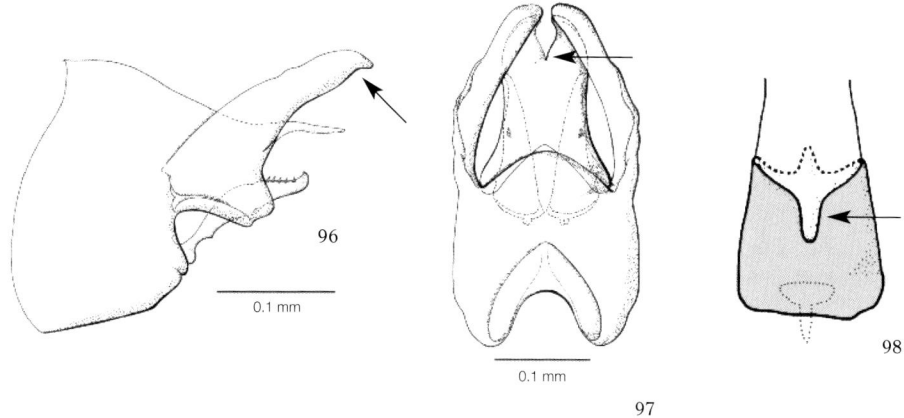

Figures 96-98. *Hydroptila sylvestris*. 96 male genitalia lateral; 97 male genitalia dorsal, aedeagus omitted; 98 female genitalia dorsal, segment VIII shaded

## *Hydroptila sparsa* **Curtis, 1834**

Fore wing length: ♂ 2.5-3 mm, ♀ 3 mm (Figs 99, 100). Common in S. England and Wales, less common in N. England and Scotland, present in Ireland; rivers and streams. Throughout Europe. Flight period: May-October, possibly bivoltine in S. England. ♂ genitalia with pointed claspers in lateral view (Fig. 101), segment X much broader than *H. sylvestris* in dorsal view (Fig. 102); ♀ genitalia with quadrate dorsal excision on segment VIII (Fig. 103), internal ventral sclerite with rounded posterior sections (Fig. 104). The separation of females from those of *H. lotensis* was not clear in Marshall's (1978) key, but was clarified by re-examination of the internal genitalic structures by Rojas-Camousseight & Tachet (1988).

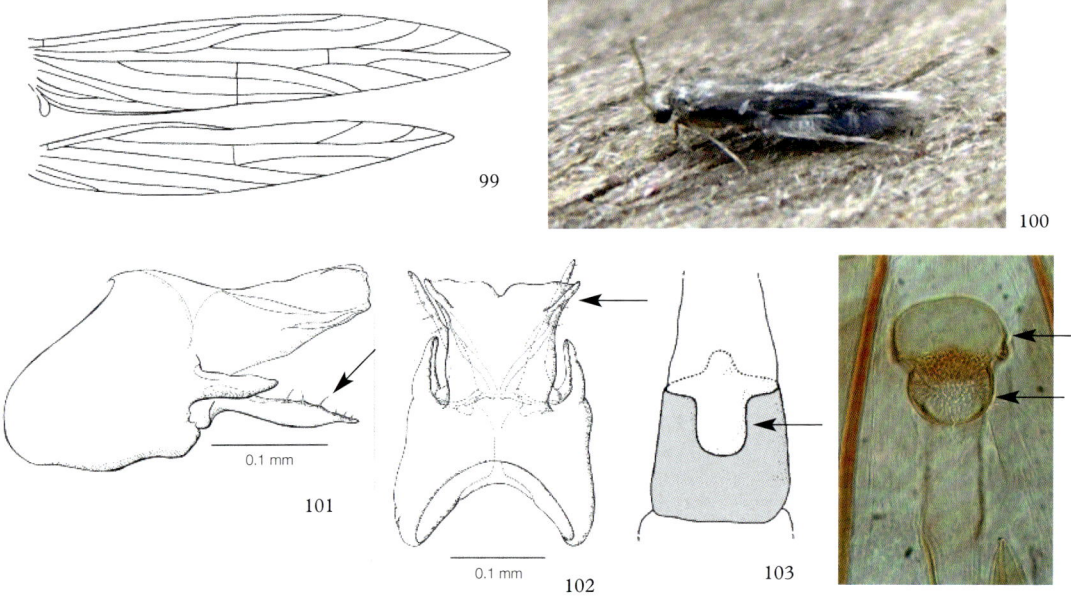

Figures 99-104. *Hydroptila sparsa*. 99 wing venation; 100 live specimen; 101 male genitalia lateral; 102 male genitalia dorsal, aedeagus omitted; 103 female genitalia dorsal, segment VIII shaded; 104 female internal ventral sclerite (slide preparation)

## *Hydroptila simulans* **Mosely, 1920**

Fore wing length: ♂ ♀ 2.5-3 mm. Throughout Britain, often common, present in Ireland; rivers and streams. Throughout much of Europe. Flight period: May-October. ♂ genitalia with claspers broadened apically in lateral view (Fig. 105), tergite X narrower in dorsal view than *H. sparsa* (Fig. 106); ♀ genitalia with very shallow dorsal excision on segment VIII (Fig. 107).

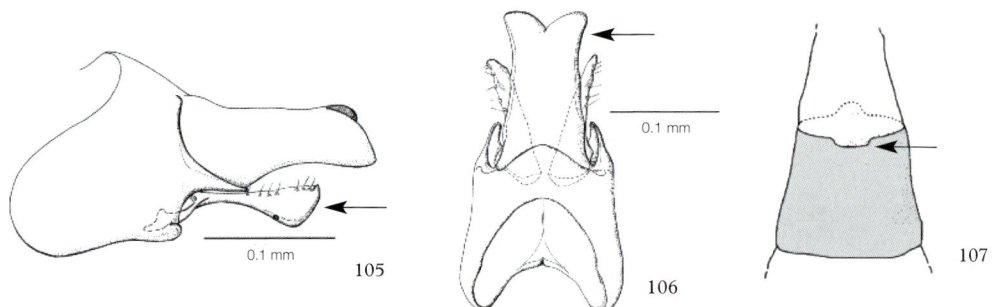

Figures 105-107. *Hydroptila simulans*. 105 male genitalia lateral;
106 male genitalia dorsal, aedeagus omitted; 107 female genitalia dorsal, segment VIII shaded

## *Hydroptila cornuta* Mosely, 1922

Fore wing length: ♂ ♀ 3 mm. SE. England, East Anglia and Shetland, not common; present in Ireland; rivers and streams. Scattered records throughout Europe, commoner in north. Flight period: May-September. ♂ genitalia with claspers similar to *H. simulans* but with a dark upturned point at the apex (Fig. 108), segment X broad with extended posterior corners in dorsal view (Fig. 109); ♀ genitalia with two shallow posterior excisions on segment VIII (Fig. 110).

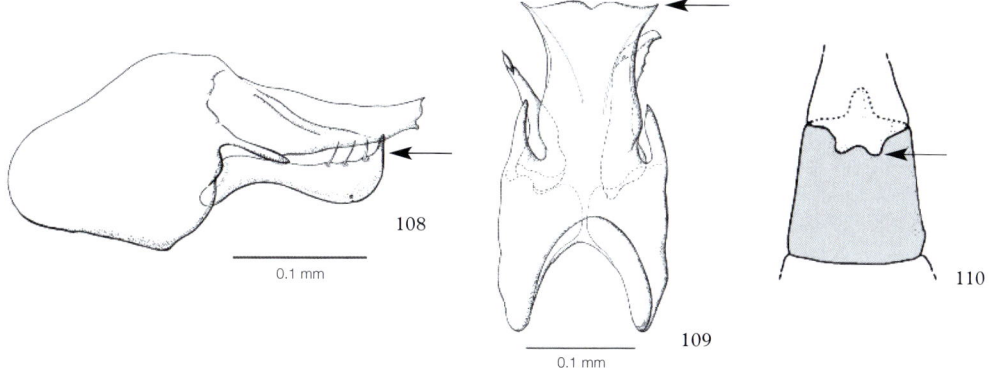

Figures 108-110. *Hydroptila cornuta*. 108 male genitalia lateral; 109 male genitalia dorsal, aedeagus omitted; 110 female genitalia dorsal, segment VIII shaded

## *Hydroptila lotensis* Mosely, 1930

This species was not recognised as British until Kimmins (1961). Fore wing length: ♂ ♀ 2.5-3 mm. Only known from the Rivers Wye, Lugg and Severn; no records from Ireland; rivers. Scattered records through western and central Europe. Flight period: June-August. ♂ genitalia with strong dorsally directed spine at apex of claspers (Fig. 111), segment X approximately parallel-sided with narrow central excision in dorso-ventral view (Fig. 112). Note that the caption to Fig. 19 in Marshall (1978) should read ventral, not dorsal. ♀ genitalia with square notch on posterior margin of segment VIII as in *H. sparsa* but internal ventral sclerite with more flattened posterior parts (Fig. 113). The separation of females in this group was clarified by Rojas-Camousseight & Tachet (1988).

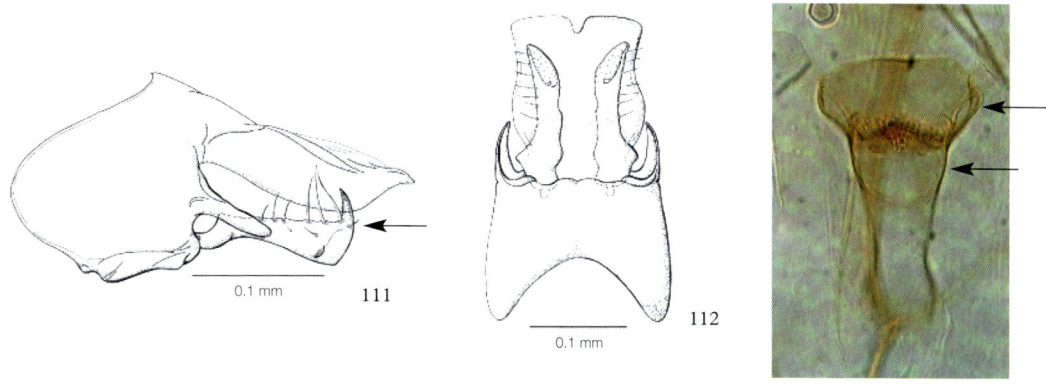

Figures 111-113. *Hydroptila lotensis*. 111 male genitalia lateral; 112 male genitalia ventral, aedeagus omitted; 113 female internal ventral sclerite (slide preparation)

## *Hydroptila angulata* **Mosely, 1922**

Fore wing length: ♂♀ 3 mm. Scotland, Wales, N. and SW. England, not common; present in Ireland; rivers. Throughout much of Europe. Flight period: June-August (October). ♂ genitalia with upturned and out-turned apices to claspers (Fig. 114), segment X with very rounded posterior margin and shallow rounded central excision (Fig. 115); ♀ genitalia with a uniquely thickened rim to the dorsal notch on segment VIII (Fig. 116).

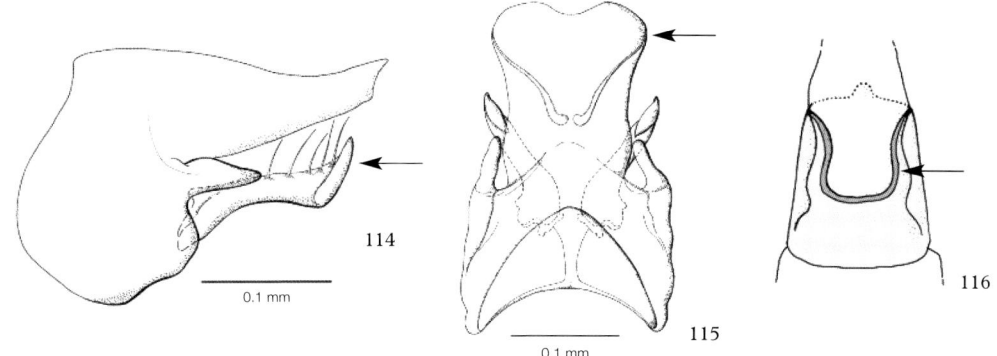

Figures 114-116. *Hydroptila angulata.* 114 male genitalia lateral;
115 male genitalia dorsal, aedeagus omitted; 116 female genitalia dorsal, segment VIII shaded

## *Hydroptila occulta* **(Eaton, 1873)**

Fore wing length: ♂♀ 2.5-3 mm. Throughout Britain, not common; present in Ireland; fast streams and rivers. Throughout much of Europe. Flight period: June-October. ♂ genitalia with short area of sclerotisation on dorsal apex of claspers (Fig. 117), apex of segment X with shallow excision, sometimes with small central process, in dorsal view (Fig. 118); ♀ genitalia with broad T-shaped ventral sclerite on sternite VIII (Fig. 119).

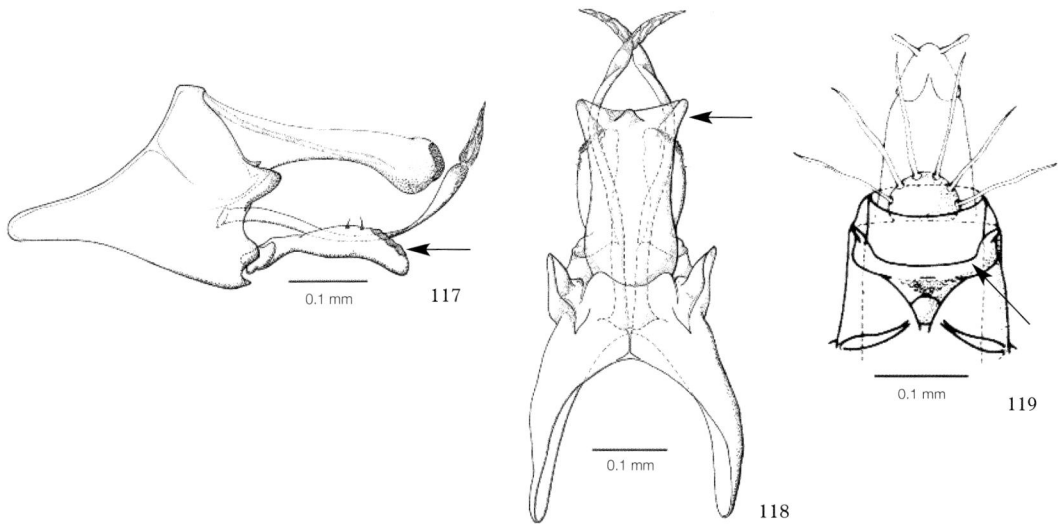

Figures 117-119. *Hydroptila occulta.* 117 male genitalia lateral;
118 male genitalia dorsal, aedeagus omitted; 119 female genitalia ventral

## *Hydroptila martini* **Marshall, 1977**

This species was added to the British list when first described by Marshall (1977). Fore wing length: ♂♀ 2.5-3 mm. S. England, Wales and SW. Scotland, not common; present in Ireland; streams and rivers. Scattered records throughout Europe except extreme north. Some early records of *H. occulta* may refer to this species. Flight period: May-October, possibly multivoltine (Cooling, 1982). ♂ genitalia with area of sclerotisation on dorsal apex of claspers extending to the clasper tip (Fig. 120), apex of segment X with central process longer than corners (Fig. 121); ♀ genitalia with narrow Y-shaped ventral sclerite on sternite VIII (Fig. 122).

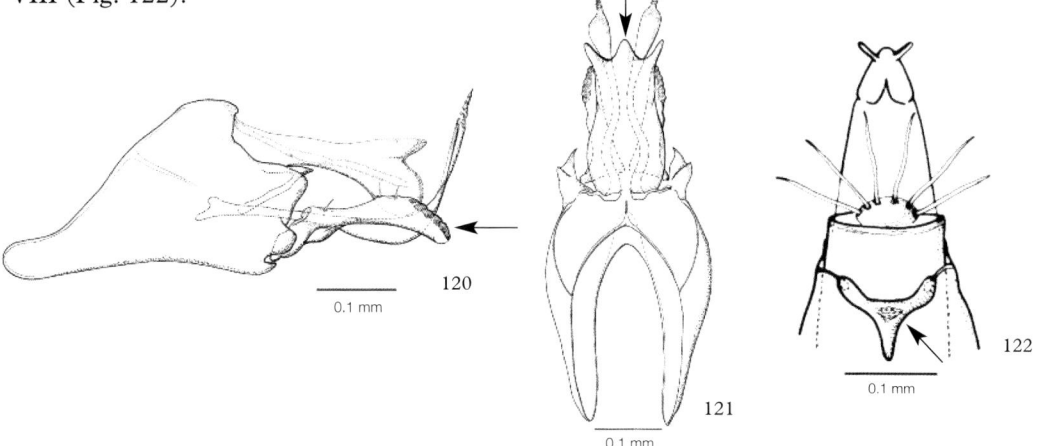

Figures 120-122. *Hydroptila martini.* 120 male genitalia lateral;
121 male genitalia dorsal, aedeagus omitted; 122 female genitalia ventral

## *Hydroptila valesiaca* **Schmid, 1947**

This species was not recognised as British until Marshall (1977). Fore wing length: ♂♀ 2.5-3 mm. Scotland, not common; no records from Ireland; spring streams. Few records, mainly central Europe. Flight period: June-July. ♂ genitalia with area of sclerotisation on dorsal apex of claspers extending much further anteriorly than in *H. martini* (Fig. 123), apex of segment X with large well-defined notch (Fig. 124); ♀ genitalia with broad Y-shaped ventral sclerite on sternite VIII, with swollen apices on the 'arms' (Fig. 125).

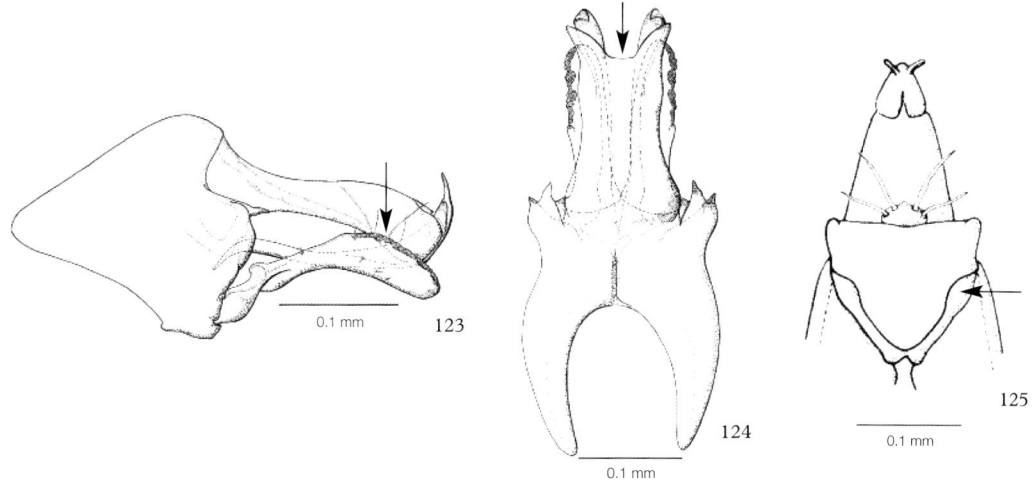

Figures 123-125. *Hydroptila valesiaca.* 123 male genitalia lateral;
124 male genitalia dorsal, aedeagus omitted; 125 female genitalia ventral

# Genus ITHYTRICHIA Eaton, 1873

Spur formula 0.3.4. Ocelli present. Of the four European species just two occur in Britain.

## Ithytrichia lamellaris Eaton, 1873

Fore wing length: ♂♀ 3-3.5 mm (Fig. 126). Common throughout Britain; present in Ireland; streams and rivers. Throughout Europe. Flight period: May-October, possibly multivoltine (Cooling, 1982). ♂ genitalia with pair of dark internal rods in segement IX, claspers broad with blunt apices (Fig. 127); ♀ genitalia with three-lobed posterior margin on sternite VIII (Fig. 128).

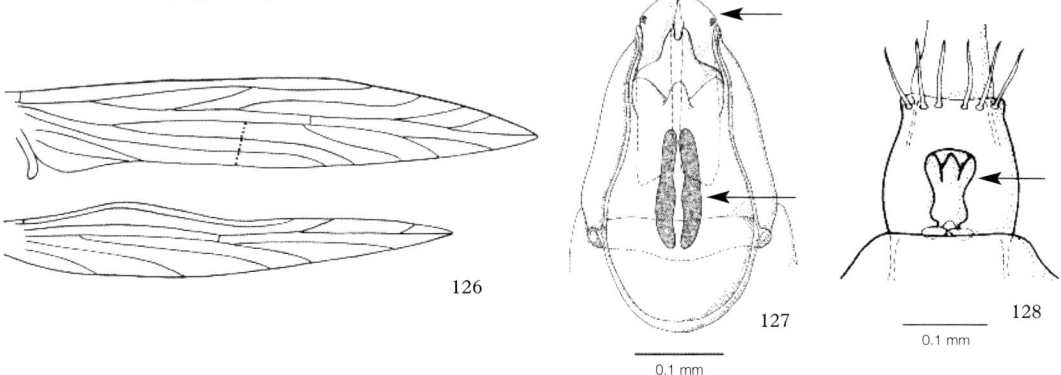

Figures 126-128. *Ithytrichia lamellaris.* 126 wing venation; 127 male genitalia dorsal, aedeagus omitted; 128 female genitalia ventral

## Ithytrichia clavata Morton, 1905

This species was not recorded as British until Grensted (1939), too late for inclusion by Mosely (1939). Fore wing as in *I. lamellaris*. Fore wing length: ♂ 2-3 mm ♀ 1.5-3 mm. Lake District, NW. Wales and S. England, but very few records; present in Ireland; fast rivers. Scattered records, mainly western Europe. Flight period: June-August. ♂ genitalia lacking internal rods, claspers narrow with pointed apices (Fig. 129); ♀ genitalia with sternite VIII single-lobed (Fig. 130). Note the error in Marshall's (1978) key to *Ithytrichia* (p. 22) where the second half of couplet 3 should read 'convex posterior margin'.

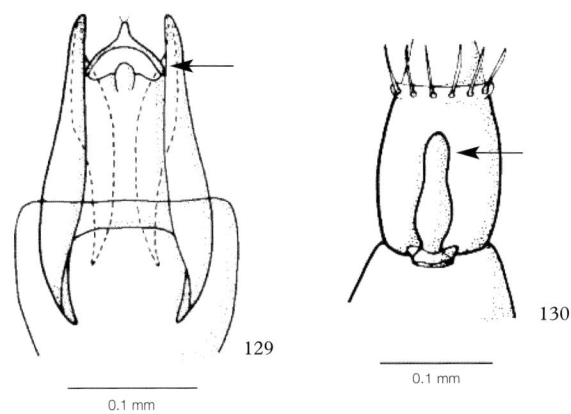

Figures 129-130. *Ithytrichia clavata.* 129 male genitalia dorsal, aedeagus omitted; 130 female genitalia ventral

# Genus AGRAYLEA Curtis, 1834

Spur formula 0.3.4. Ocelli present. Both European species are found in Britain.

## *Agraylea multipunctata* **Curtis, 1834**

Fore wing length: ♂ 2.5-4 mm ♀ 3.5-4 mm (Figs 131, 132). Common throughout Britain; present in Ireland; weedy ponds or slow rivers. Throughout Europe, less common in south. Flight period: May-September, possibly bivoltine. ♂ genitalia with small L-shaped claspers with narrow apices (Fig. 133); ♀ genitalia with sternite VIII large and conspicuous with thickened margins, internal apparatus elongate (Fig. 134).

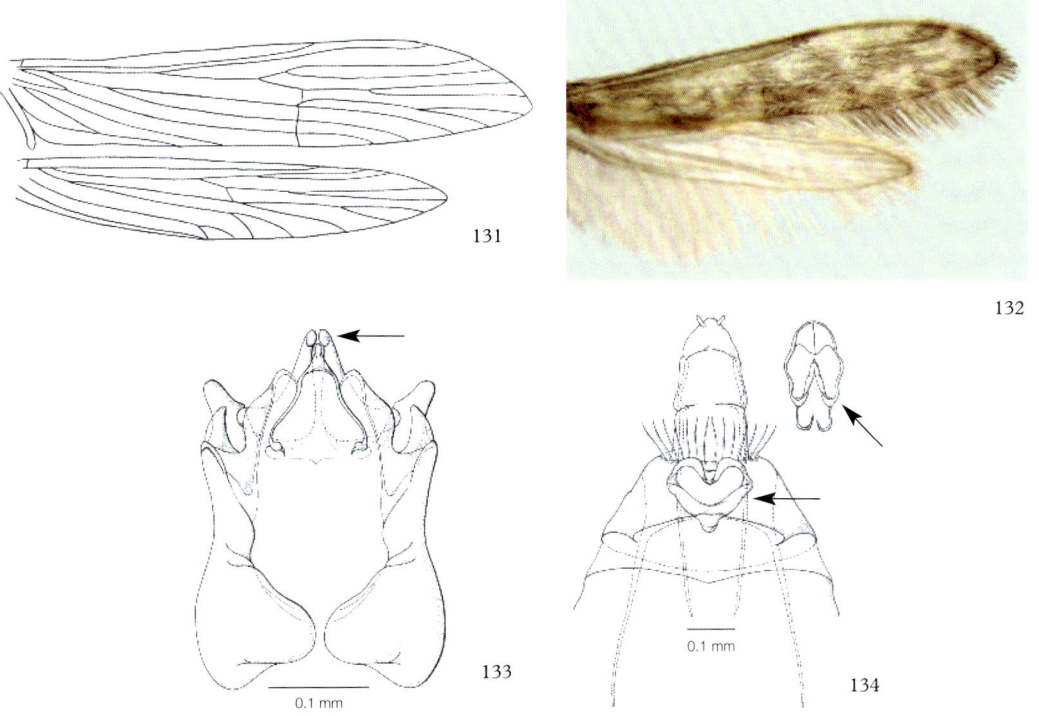

Figures 131-134. *Agraylea multipunctata*. 131 wing venation; 132 wing pattern;
133 male genitalia dorsal, aedeagus omitted; 134 female genitalia ventral, internal structure inset

### *Agraylea sexmaculata* **Curtis, 1834**
= *pallidula* McLachlan, 1875

Listed as *A. pallidula* in Mosely (1939). Fore wing length: ♂ 3-4.5 mm ♀ 2.5-3.5 mm. Throughout Britain, but apparently less common than *A. multipunctata*; present in Ireland; weedy ponds or slow rivers. Throughout Europe. Flight period: May-September. ♂ genitalia with large and broad claspers with darkened rim (Fig. 135); ♀ genitalia with sternite VIII small and inconspicuous, internal apparatus much shorter and broader than *A. multipunctata* (Fig. 136).

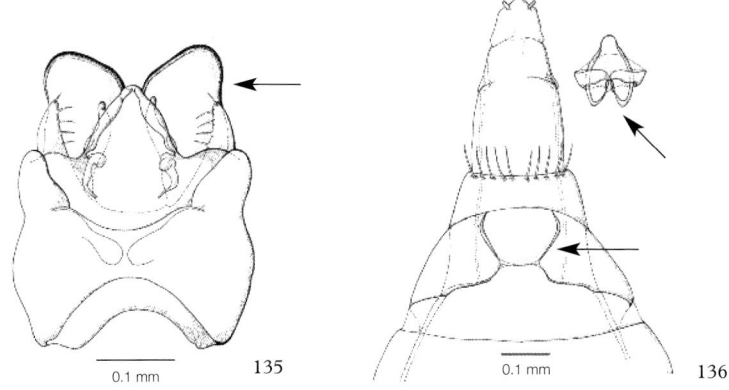

Figures 135-136. *Agraylea sexmaculata.* 135 male genitalia dorsal, aedeagus omitted; 136 female genitalia ventral, internal structure inset

## Genus ALLOTRICHIA McLachlan, 1880

Spur formula 0.3.4. Ocelli present. Of about six European species, just one occurs in Britain.

### *Allotrichia pallicornis* **(Eaton, 1873)**

Fore wing length: ♂ ♀ 4-5 mm (Fig. 137). Widespread throughout Britain but not common; present in Ireland; fast, stony streams and rivers. Throughout Europe, except north. Flight period: June-August. ♂ genitalia with broad claspers with slightly concave posterior margins (Fig. 138); ♀ genitalia with an asymmetrical groove on sternite VIII, internal apparatus very long (Fig. 139).

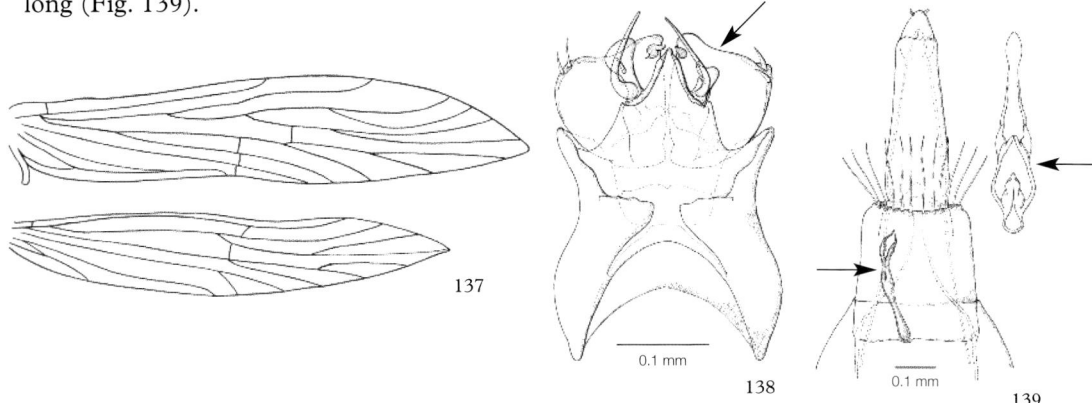

Figures 137-139. *Allotrichia pallicornis.* 137 wing venation; 138 male genitalia dorsal, aedeagus omitted; 139 female genitalia ventral, internal structure inset

# Genus ORTHOTRICHIA Eaton, 1873

Spur formula 0.3.4. Ocelli absent. Of the eight European species three occur in Britain. All species have asymmetrical male genitalia.

## *Orthotrichia angustella* (McLachlan, 1865)

Fore wing with white fringes that form a distinctive white stripe when the wings are folded. Fore wing length: ♂♀ 2.5-3 mm (Fig. 140). The male has a short row of dark scent-scales along the base of the subcosta in the fore wing (Fig. 141). Present in England, but few records; present in Ireland; ponds, lakes and slow rivers. Throughout much of Europe. Flight period: June-August. ♂ genitalia with a pair of well-developed processes on segment IX, of similar size (Fig. 142); ♀ genitalia with slightly rounded posterior margin to segment VIII in lateral view (Fig. 143), with rounded apical excision in posterior dorsal margin of segment VIII and oblique ventral groove on ventral side (Fig. 144). Note the error in couplet 5 of Marshall's (1978) key to females of *Orthotrichia*: the first half of the key should read "(Fig. 45) *costalis*" and the second half "(Fig. 43) *angustella*".

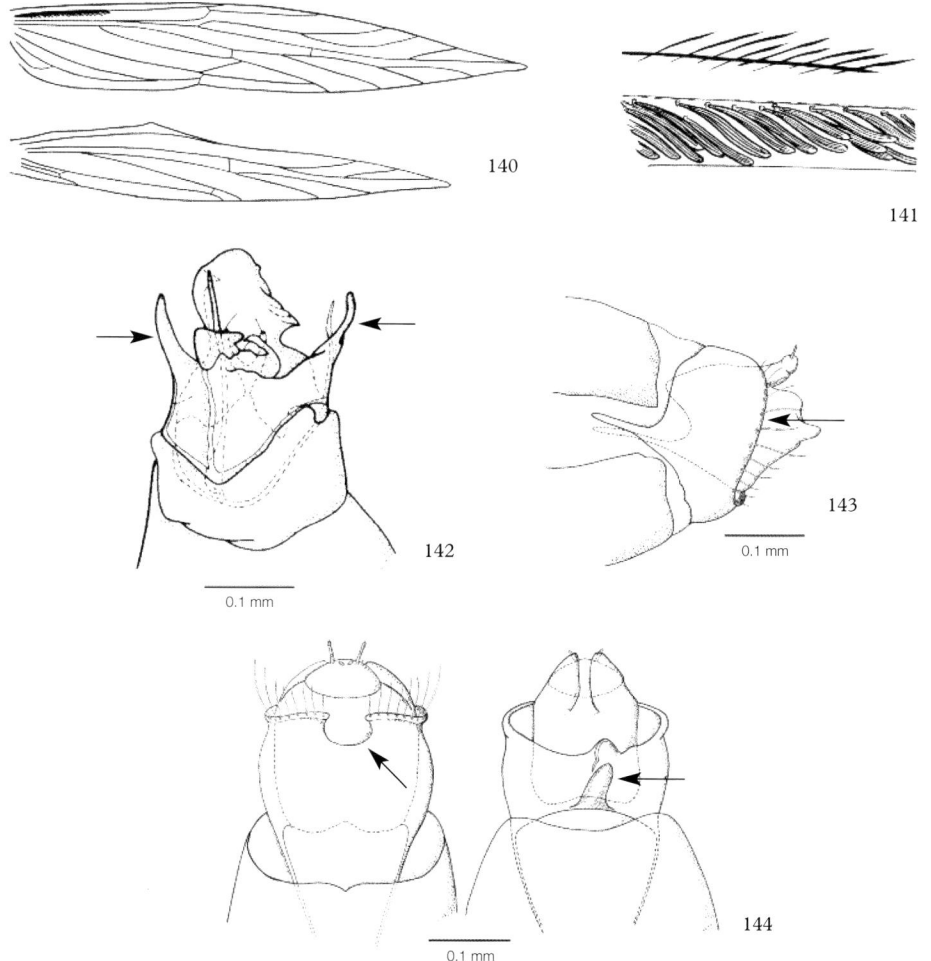

Figures 140-144. *Orthotrichia angustella*. 140 wing venation; 141 male fore wing subcostal scent scales; 142 male genitalia ventral, aedeagus omitted; 143 female genitalia lateral; 144 female genitalia dorsal and ventral

## *Orthotrichia costalis* (Curtis, 1834)
= *tetensii* Kolbe, 1887

Listed as *O. tetensii* in Mosely (1939). Fore wing as in *O. angustella*, with white fringes forming distinctive white stripe when the wings are folded. The male has no row of scent-scales on the subcosta. Fore wing length: ♂♀ 3 mm (Fig. 145). Throughout Britain, but not common; present in Ireland; still or slow water. Throughout much of Europe. Flight period: May-September. ♂ genitalia with single process on segment IX on right side only, claspers large and incurved (Fig. 146); ♀ genitalia with straight posterior margin to segment VIII in lateral view (Fig. 147), no apical excision in posterior margin, but with a pair of dark triangular sclerites ventrally (Fig. 148). See the corrections to Marshall's key, under *O. angustella* (above).

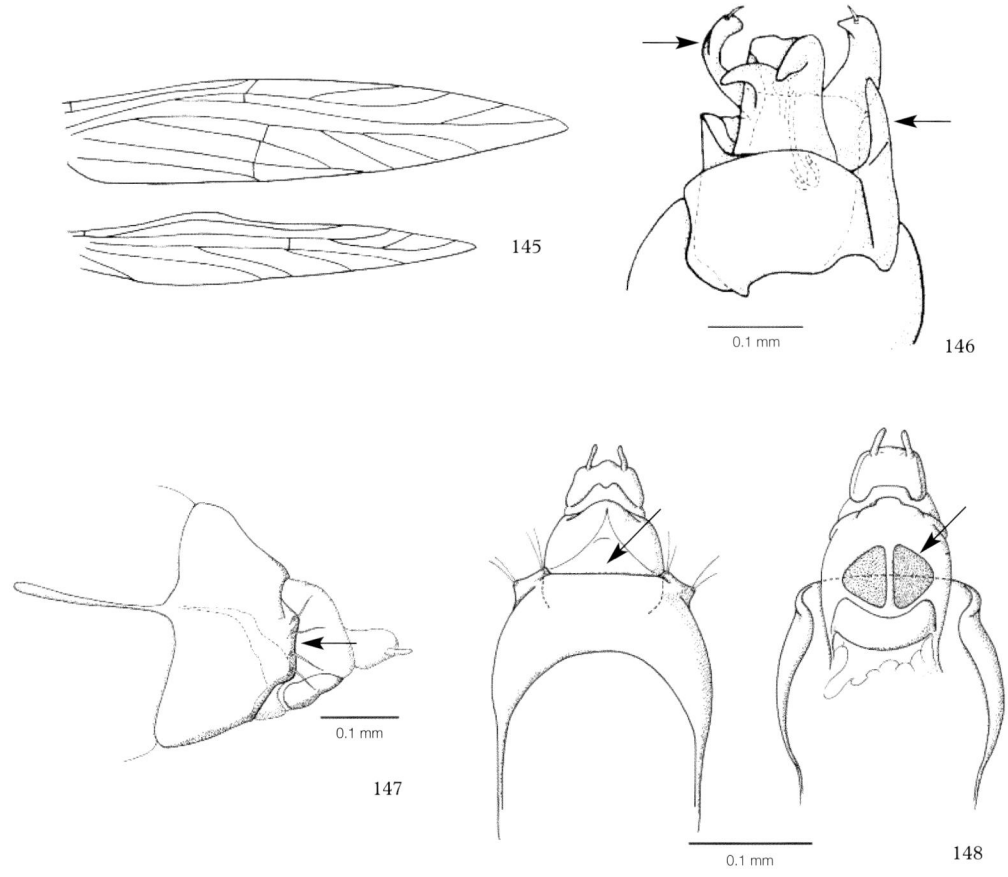

Figures 145-148 *Orthotrichia costalis*. 145 wing venation; 146 male genitalia dorsal, aedeagus omitted; 147 female genitalia lateral; 148 female genitalia dorsal and ventral

## *Orthotrichia tragetti* **Mosely, 1930**

Fore wing length: ♂♀ 2.5-3 mm, similar to *O. costalis*. One early record from Hampshire; probably an extinct introduction; no records from Ireland. Scattered records throughout Europe. Flight period: July. ♂ genitalia with no processes on segment IX, claspers small and inconspicuous (Fig. 149); ♀ genitalia with slender ventral process, conspicuous in both lateral (Fig. 150) and ventral (Fig. 151) views.

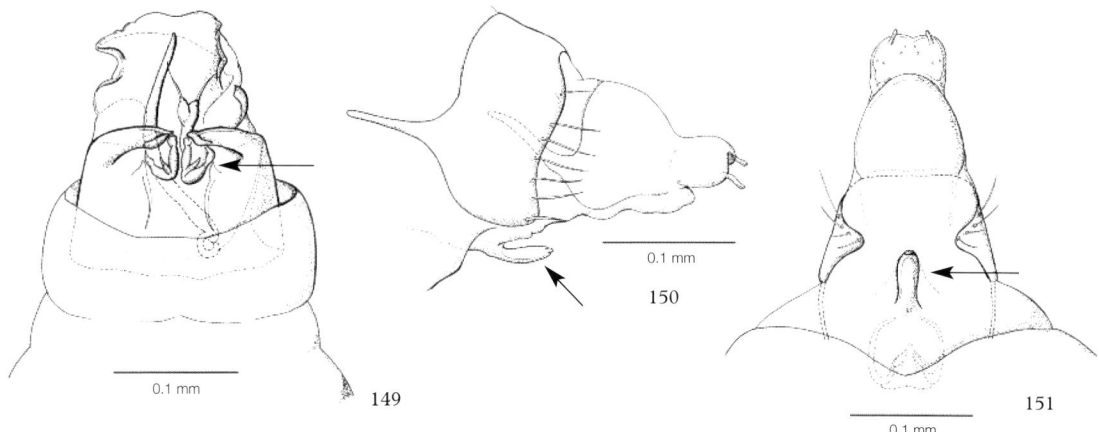

Figures 149-151 *Orthotrichia tragetti*. 149 male genitalia dorsal, aedeagus omitted;
150 female genitalia lateral; 151 female genitalia ventral

## Genus TRICHOLEIOCHITON Kloet & Hincks, 1944

Spur formula 0.3.4. Ocelli present. *Tricholeiochiton fagesii* is the only European species of this genus.

## *Tricholeiochiton fagesii* **(Guinard, 1879)**

Listed as *Leiochiton fagesii* in Mosely (1939). Fore wing length: ♂♀ 3 mm (Fig. 152). Scotland, Wales, NW. and S. England, but very few records; present in Ireland; weedy ponds and lakes. Scattered records throughout Europe. Flight period: July. ♂ genitalia with distinctive square shape (Fig. 153); ♀ genitalia with large ventral lobe on segment VIII (Fig. 154).

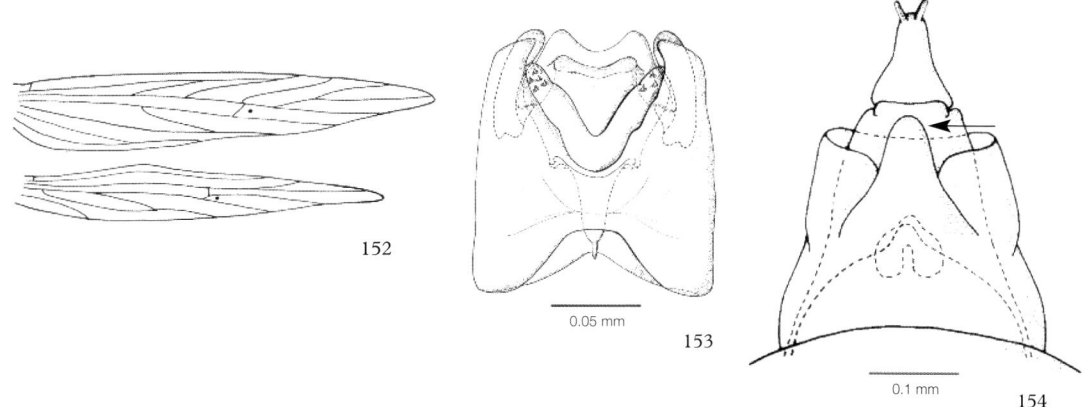

Figures 152-154 *Tricholeiochiton fagesii*. 152 wing venation;
153 male genitalia ventral, aedeagus omitted; 154 female genitalia ventral

# Genus OXYETHIRA Eaton, 1873

Spur formula 0.3.4. Ocelli present. Of around 20 European species eight are found in Britain.

## *Oxyethira frici* **Klapálek, 1891**

Fore wing length: ♂♀ 2 mm. Scotland, Wales, NW. and SW. England, not common; present in Ireland; streams and rivers. Scattered records throughout Europe. Flight period: May-September. ♂ genitalia with long dark spines on segment VIII (Fig. 155); ♀ genitalia with large subtriangular posterior lobes on internal apparatus (Fig. 156).

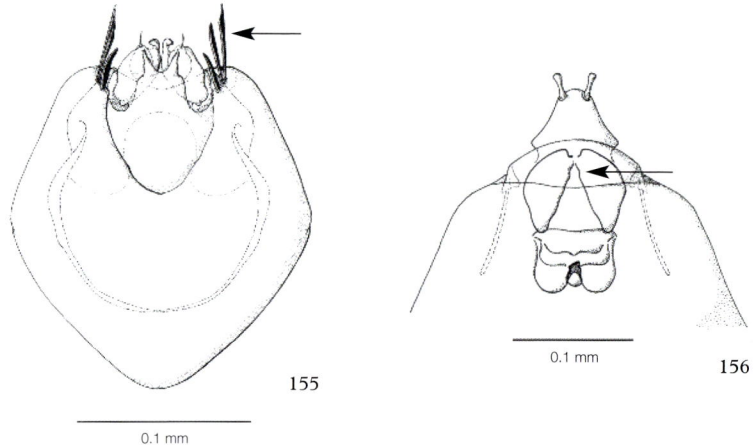

Figures 155-156. *Oxyethira frici*. 155 male genitalia ventral, aedeagus omitted; 156 female genitalia ventral

## *Oxyethira flavicornis* **(Pictet, 1834)**
= *costalis* Eaton

Listed as *O. costalis* in Mosely (1939). Fore wing length: ♂♀ 3-3.5 mm (Figs 157, 158). Common throughout Britain; present in Ireland; ponds and lakes. Scattered records throughout much of Europe. Flight period: May-September, possibly bivoltine. ♂ genitalia with hooked lateral margins of segment VIII (Fig. 159); ♀ genitalia with strongly concave lateral margins of fused segments VIII+IX (Fig. 160).

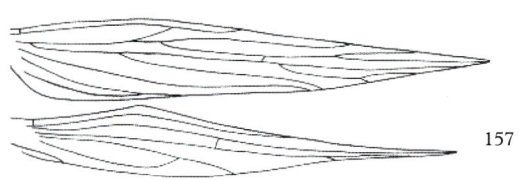

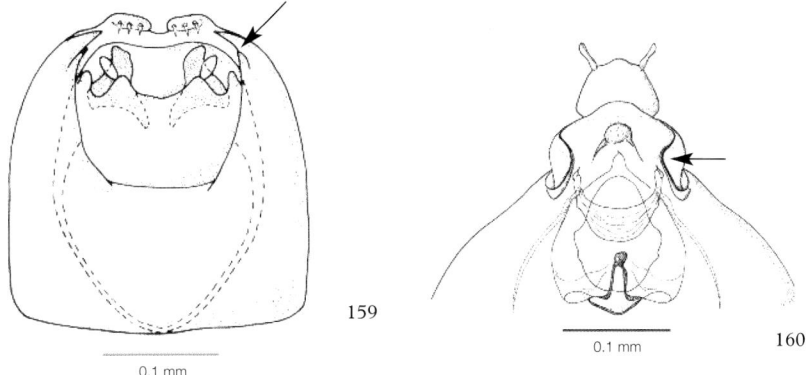

Figures 157-160. *Oxyethira flavicornis*. 157 wing venation; 158 wing pattern; 159 male genitalia ventral, aedeagus omitted; 160 female genitalia dorsal

## *Oxyethira simplex* Ris, 1897

Fore wing length: ♂♀ 2.5 mm. Scattered throughout Britain, but not common; present in Ireland; streams, rivers and lakes. Scattered records throughout Europe. Flight period: April-September. ♂ genitalia rounded and setose, but with no hooks or spines (Fig. 161); ♀ genitalia with narrow posterior sclerites that converge apically in a bell-shape (Fig. 162).

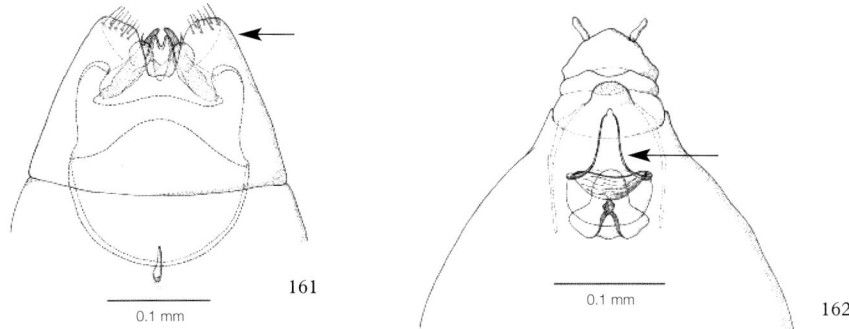

Figures 161-162. *Oxyethira simplex*. 161 male genitalia ventral, aedeagus omitted; 162 female genitalia dorsal

## *Oxyethira tristella* Klapálek, 1895

Fore wing length: ♂♀ 3 mm. Scattered throughout Britain but no recent records; present in Ireland; streams, rivers and lakes. Scattered records through central and northern Europe. Flight period: May-September. ♂ genitalia with sharply inturned processes on segment IX (Fig. 163); ♀ genitalia with dark heart-shaped patch on sternum VIII (Fig. 164).

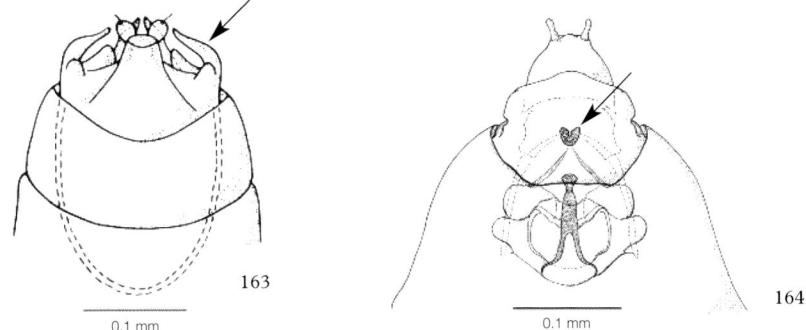

Figures 163-164. *Oxyethira tristella*. 163 male genitalia ventral, aedeagus omitted; 164 female genitalia ventral

## *Oxyethira falcata* Morton, 1893

Fore wing length: ♂♀ 3.5 mm. Common throughout Britain; present in Ireland; streams, rivers and lakes. Throughout much of Europe. Flight period: (February-) May-November (December), possibly multivoltine (Cooling, 1982). ♂ genitalia with short dark spines on segment VIII (Fig. 165), much shorter than those in *O. frici* (cf. Fig. 155); ♀ genitalia with dark comma-shaped posterior lobes on internal apparatus (Fig. 166).

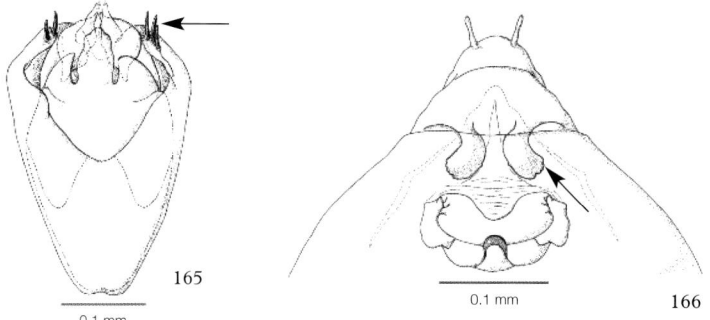

Figures 165-166. *Oxyethira falcata.* 165 male genitalia ventral, aedeagus omitted; 166 female genitalia ventral

## *Oxyethira mirabilis* Morton, 1904

Listed as *Oxytrichia mirabilis* in Mosely (1939). Although Mosely decided that *O. mirabilis* deserved a new genus, the venational and genitalic characters that he cited are no longer accepted as indicative of a distinct genus. When Kimmins (1966) transferred *O. mirabilis* (the type species of *Oxytrichia*) into *Oxyethira* he effectively synonymised the two genera. Fore wing length: ♂♀ 2.5 mm (Fig. 167). Few records, but now known from Wales, Scotland and northern England; no records from Ireland; flowing water in upland mires and bogs. Few records in northern Europe only. Flight period: July. ♂ genitalia very simplified, with few processes (Figs 168, 169); ♀ genitalia with extremely long internal apparatus (Fig. 170).

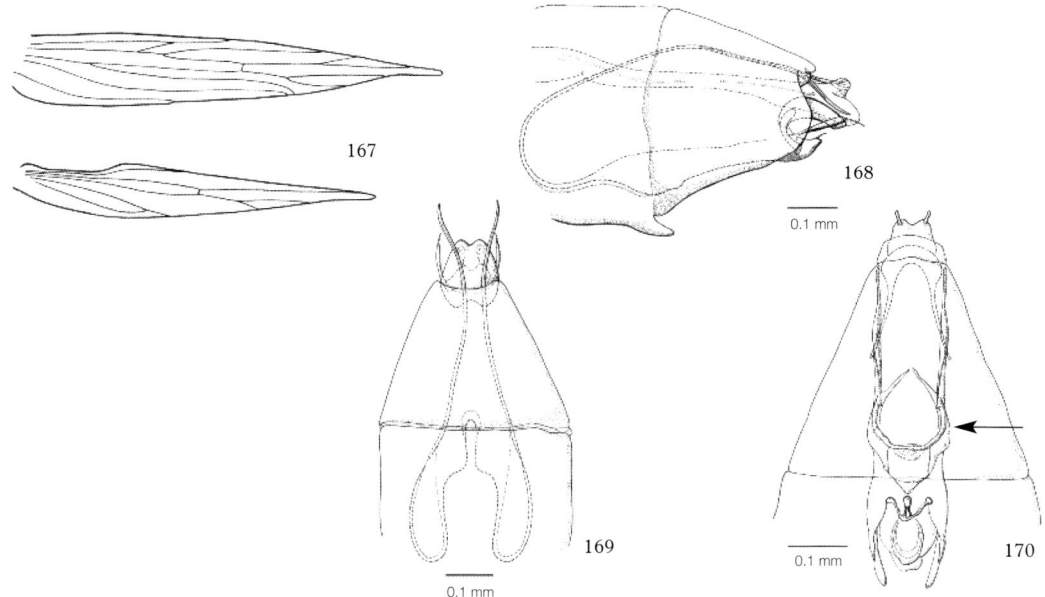

Figures 167-170. *Oxyethira mirabilis.* 167 wing venation; 168 male genitalia lateral;
169 male genitalia dorsal, aedeagus omitted; 170 female genitalia ventral

## *Oxyethira sagittifera* **Ris, 1897**

Fore wing length: ♂♀ 2.5 mm. Scotland and NW. England, not common; present in Ireland; lakes and large ponds. Few records in northern Europe only. Flight period: June-August. ♂ genitalia with dark sinuate ventral appendages (Fig. 171); ♀ genitalia with broad V-shaped excision on posterior margin of sternite VIII (Fig. 172).

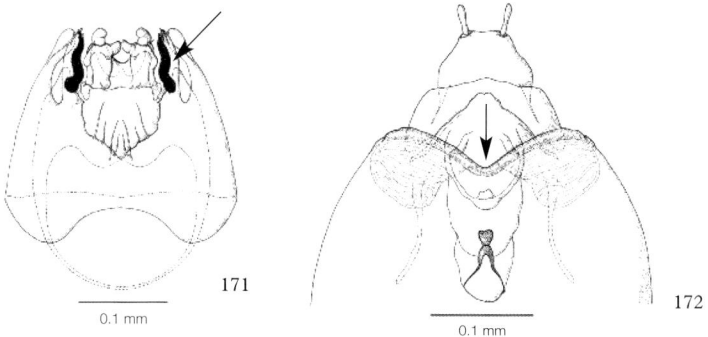

Figures 171-172. *Oxyethira sagittifera.* 171 male genitalia dorsal, aedeagus omitted; 172 female genitalia ventral

## *Oxyethira distinctella* **McLachlan, 1880**

Fore wing length: ♂♀ 2.5 mm. One early record from Hampshire; probably an extinct introduction; no records from Ireland. Northern Europe only. Flight period: July. ♂ genitalia with elongate segment VIII, bearing long slender processes that extend beyond segment IX (Fig. 173); ♀ genitalia with deep V-shaped posterior excision on sternite VIII (Fig. 174).

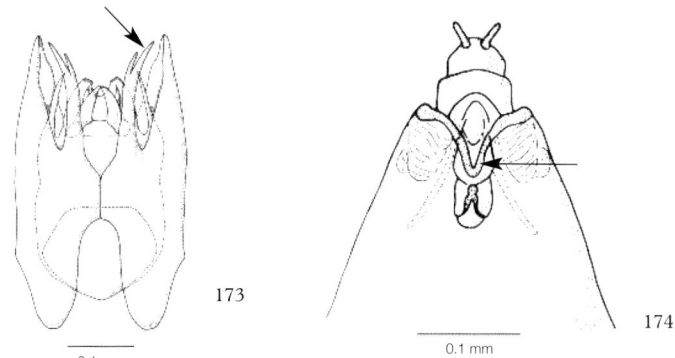

Figures 173-174. *Oxyethira distinctella.* 173 male genitalia dorsal, aedeagus omitted; 174 female genitalia ventral

# Family PHILOPOTAMIDAE (3 genera, 5 species)

Fore wing with forks 1, 2, 3 and 5 (fork 4 present in *Philopotamus*), hind wing with forks 1, 2, 3 and 5; discoidal cell closed in both wings. Spur formula 1.4.4 or 2.4.4. Ocelli present.

## Key to genera of Philopotamidae

1. Spur formula 1.4.4, fore wing with yellow streaks along front and hind margins ..............
..................................................................................... *Chimarra* (p. 54)

- Spur formula 2.4.4, fore wings spotted or plain coloured ........................................... 2


2. Fork 4 present in fore wing (Fig. 175); fore wing strongly spotted ..... *Philopotamus* (p. 55)

- Fork 4 absent in fore wing (Fig. 176); fore wing plain .......................... *Wormaldia* (p. 57)

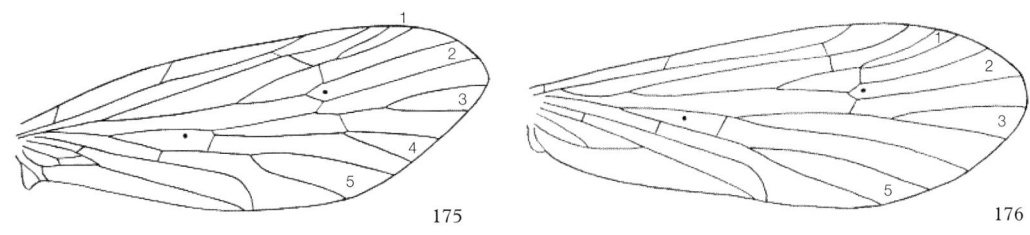

175                                        176

# Genus CHIMARRA Stephens, 1829

Fore wing with forks 1, 2, 3 and 5 (Fig. 177). Spur formula 1.4.4. There are around six European species, with just one occurring in Britain.

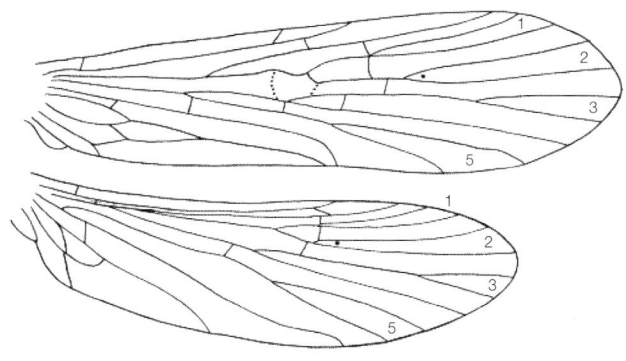

Figure 177. *Chimarra marginata*. Wing venation

## *Chimarra marginata* (Linnaeus, 1761)

Listed as *Chimarrha* [sic] *marginata* in Mosely (1939). Fore wing length: ♂ 5-7 mm, ♀ 6-8 mm (Figs 177, 178). The yellow streaks on the fore wing, and along the costal margin of the hind wing, are very conspicuous in life. A very local species in S., SW. and N. England, Wales and the bordering counties of England, mainland Scotland; present in Ireland; fast rivers. Throughout most of Europe. Flight period: April-September. ♂ genitalia with superior appendages divided into small dorsal and larger ventral branches, claspers large but simple (Fig. 179); ♀ genitalia with characteristic segment IX, L-shaped in lateral view (Fig. 180).

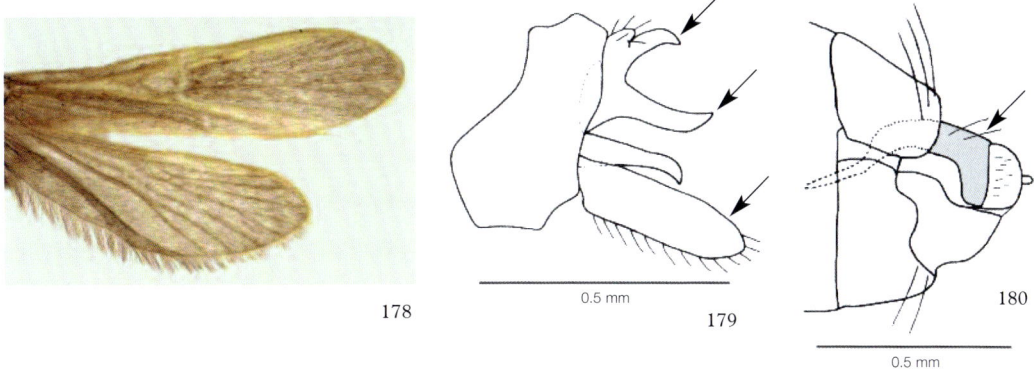

Figures 178-180. *Chimarra marginata*. 178 wing pattern;
179 male genitalia lateral, aedeagus omitted; 180 female genitalia lateral, segment IX shaded

# Genus PHILOPOTAMUS Stephens, 1829

Fore wing with forks 1 to 5 (Fig. 181). Spur formula 2.4.4. There are around 10 European species, with just one in Britain.

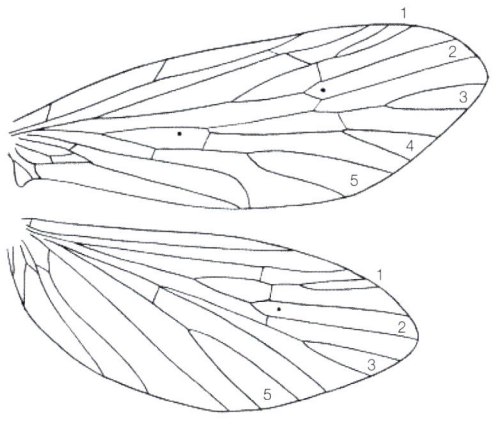

Figure 181. *Philopotamus montanus*. Wing venation

## *Philopotamus montanus* (Donovan, 1813)

= *montanus cesareus* McLachlan, 1884
= *montanus chrysopterus* Morton, 1884 (Fig. 184)
= *montanus insularis* McLachlan, 1878
= *montanus scoticus* McLachlan, 1862 (Fig. 183)

The colour forms of this variable species have been variously considered as subspecies or varieties (Mosely, 1939). They are currently regarded as colour varieties with no taxonomic status, although *cesareus* and *insularis* have been recorded only from the Channel Islands. This is the angler's Yellow Spotted Sedge. Fore wing length: ♂♀ 8-13 mm (Figs 181-185); the typical form has brown fore wings with bright yellow spots (Fig. 185). Common in SW. and N. England, Wales and bordering counties of England, mainland Scotland; present in Ireland; fast flowing streams and rivers with a stony bottom. Throughout Europe. Flight period: (March) April-October. ♂ genitalia with long claspers divided apically into downcurved dorsal branch which bears strong teeth, and straight lower branch (Fig. 186); ♀ genitalia with pale U-shaped area on dorsal side of segment IX in lateral view (Fig. 187).

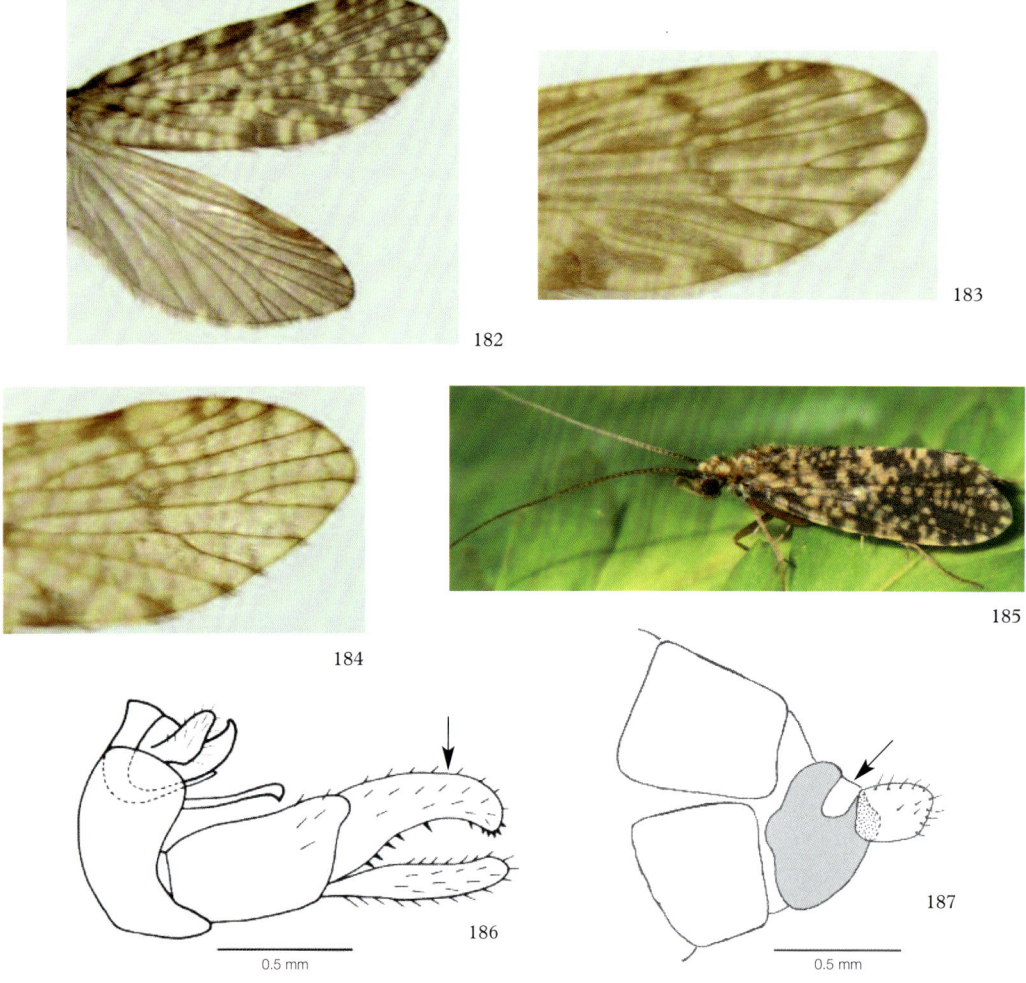

Figures 182-187. *Philopotamus montanus*. 182 wing pattern, typical; 183 form scoticus; 184 form chrysopterus; 185 live specimen; 186 male genitalia lateral, aedeagus omitted; 187 female genitalia lateral, segment IX shaded

# Genus WORMALDIA McLachlan, 1865

Fore wing with forks 1, 2, 3 and 5 (Fig. 188). Spur formula 2.4.4. A large and complex genus with around 35 European species, but the status of many of these is unclear; just three are currently recognised in Britain. At present there seems to be no reliable way of separating the females in this group and the genitalic characters described below must be treated with caution as they are based on very small samples.

## *Wormaldia occipitalis* (Pictet, 1834)

On-going work on the European populations of *Wormaldia "occipitalis"* strongly suggests that there are two species in this complex, but the nomenclature has yet to be clarified. Although there seems to be a single species in Britain this may have to be renamed in the future. Fore wing length: ♂♀ 5-7 mm (Figs 188, 189). SW. and N. England, Wales and the bordering English counties, Scotland; present in Ireland; small permanent streams and trickles. Mainly southern and central Europe. Flight period: February-March, (May) June–October. ♂ genitalia with superior appendages broader posteriorly with slightly oblique apex, apical segment of clasper broad and rounded (Fig. 190); ♀ genitalia dorsal edge of sternite IX slightly rounded in lateral view (Fig. 191).

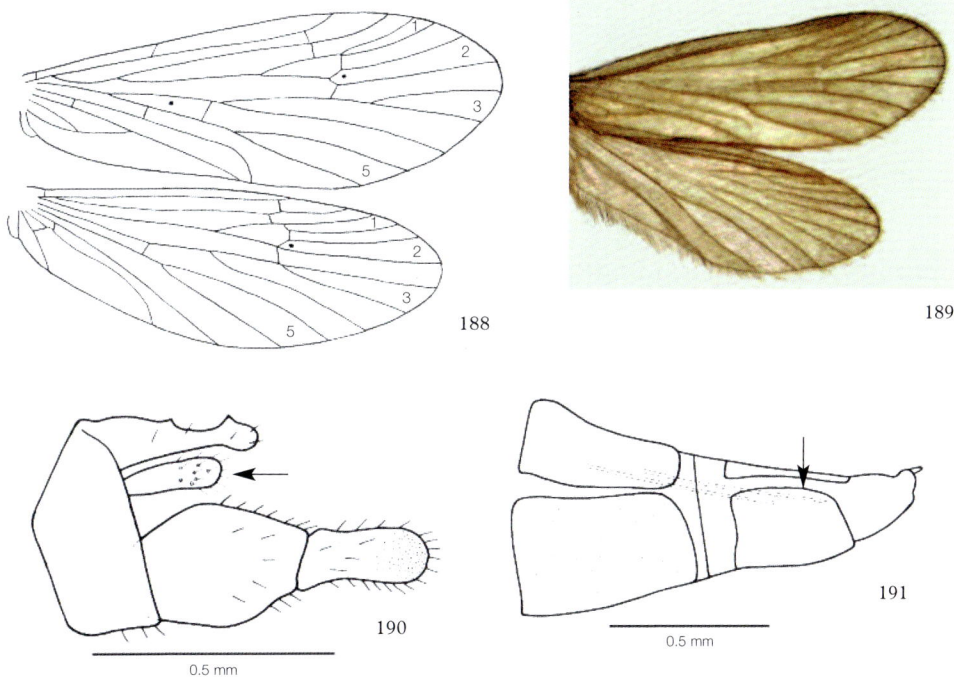

Figures 188-191. *Wormaldia occipitalis.* 188 wing venation; 189 wing pattern; 190 male genitalia lateral, aedeagus omitted; 191 female genitalia lateral

## *Wormaldia mediana* McLachlan, 1878

Synonymised with *occipitalis* by Mosely (1939) but was later recognised as a good species (Kimmins, 1953). Fore wing length: ♂ ♀ 5-7 mm (Fig. 192). A local species in SW. and N. England, Wales and Scotland; present in Ireland; large streams. Mainly southern and western Europe. Flight period: June-August. ♂ genitalia with superior appendages rounded at apex, sometimes slightly club-shaped, apical segment of clasper long and tapering (Fig. 193); ♀ genitalia very similar to those of *W. occipitalis*, but in lateral view sternite IX may appear tapered with slightly concave dorsal margin (Fig. 194).

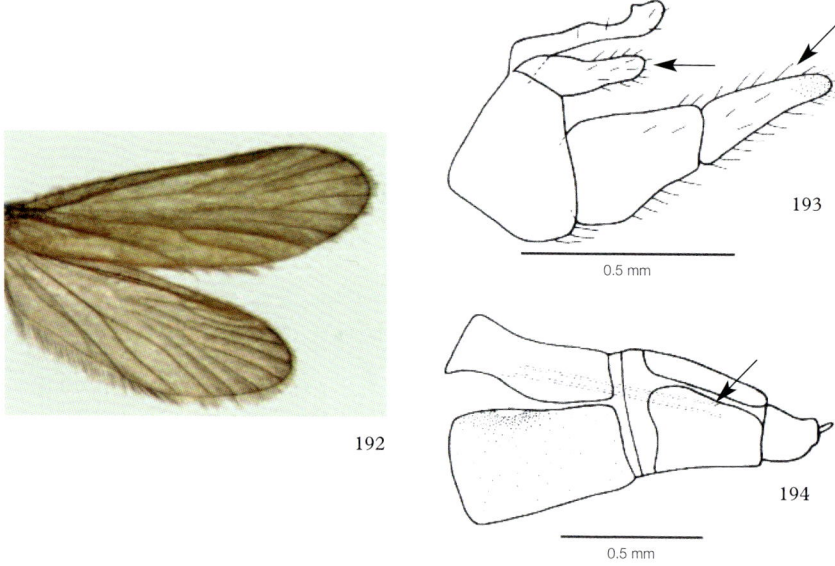

Figures 192-194. *Wormaldia mediana*. 192 wing pattern;
193 male genitalia lateral, aedeagus omitted; 194 female genitalia lateral

## *Wormaldia subnigra* McLachlan, 1865

Fore wing darker than *occipitalis* and *mediana* in life but this difference is less noticeable in older specimens (Figs 195, 196). Fore wing length: ♂♀ 5-7 mm. A local species in S. and N. England, Wales and Scotland; present in Ireland; large streams and rivers. Throughout much of Europe. Flight period: June-October. ♂ genitalia downturned end to superior appendages, apical segment of clasper narrow and tapering, but appearing shorter than in *W. mediana* (Fig. 197); ♀ genitalia very similar to those of *W. occipitalis*, but in lateral view sternite IX appears longer with a straight dorsal margin (Fig. 198).

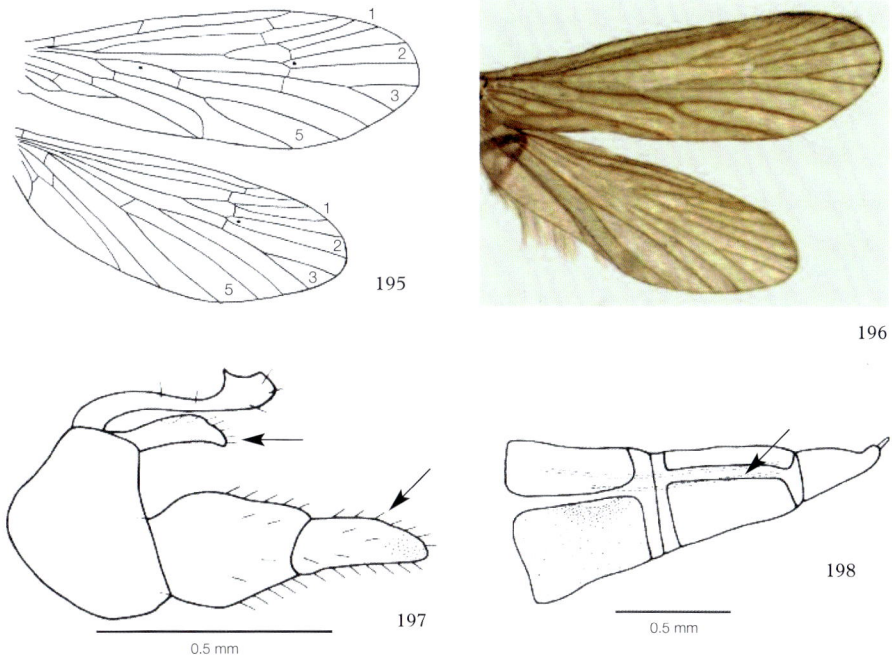

Figures 195-198. *Wormaldia subnigra*. 195 wing venation; 196 wing pattern; 197 male genitalia lateral, aedeagus omitted; 198 female genitalia lateral

# Family ECNOMIDAE (1 genus, 1 species)

Fore wing with forks 1 to 5 present, hind wing with forks 2 and 5 only; discoidal cell closed in fore wing but open in hind wing (Fig. 199). Spur formula 3.4.4. Ocelli absent. *Ecnomus tenellus* is the only British species in this family; around 10 species are known from Europe. Superficially it resembles some Polycentropodidae but the presence of a forked vein R1 in the fore wing (Fig. 199) and the absence of abdominal filaments should make separation straightforward.

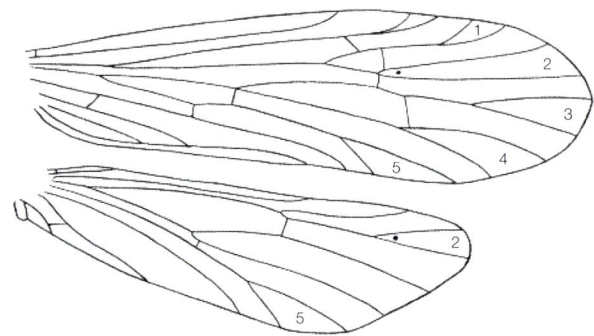

Figure 199. *Ecnomus tenellus.* Wing venation

## Genus ECNOMUS McLachlan, 1864

### *Ecnomus tenellus* (Rambur, 1842)

Fore wing length: ♂ 4-6 mm, ♀ 5-6 mm (Figs 199, 200). S.England, Midlands and southern Wales, a local species though sometimes abundant; present in Ireland; still or slow water. Throughout Europe. Flight period: June-September. ♂ genitalia (Fig. 201); ♀ genitalia (Fig. 202).

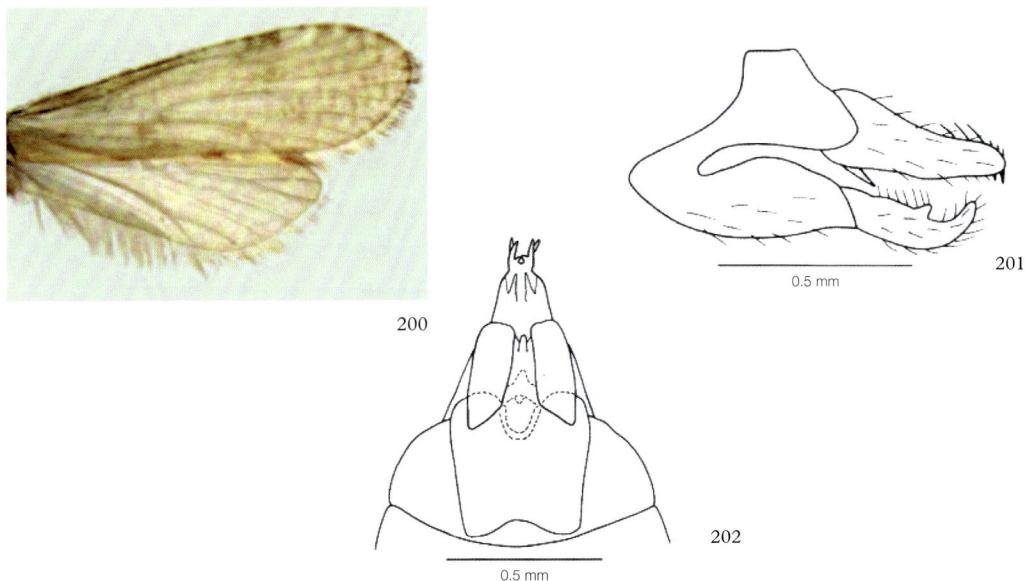

Figures 200-202. *Ecnomus tenellus.* 200 wing pattern; 201 male genitalia lateral, aedeagus omitted; 202 female genitalia ventral

# Family POLYCENTROPODIDAE (5 genera, 13 species)

Venation differs for each genus and is used to key out the genera below. Spur formula 3.4.4. Ocelli absent. Several members of this family are known as the angler's Dark Spotted Sedges. The family name was often mis-spelled as Polycentropidae by earlier authors. The identification of females in this family can sometimes be difficult and the genitalia usually need to be cleared; the ventral appendages in some genera are small and inconspicuous but they provide the best diagnostic characters. All members of this family bear a pair of slender filaments on the fifth abdominal segment, which helps to distinguish them from the Ecnomidae, but this character can be difficult to see; however, no members of the Polycentropodidae have vein R1 forked in the fore wing.

## Key to genera of Polycentropodidae

1. Apical fork 1 absent in fore wing (Fig. 203) ............................................... *Cyrnus* (p. 65)

- Apical fork 1 present in fore wing (Fig. 204) ................................................................. 2

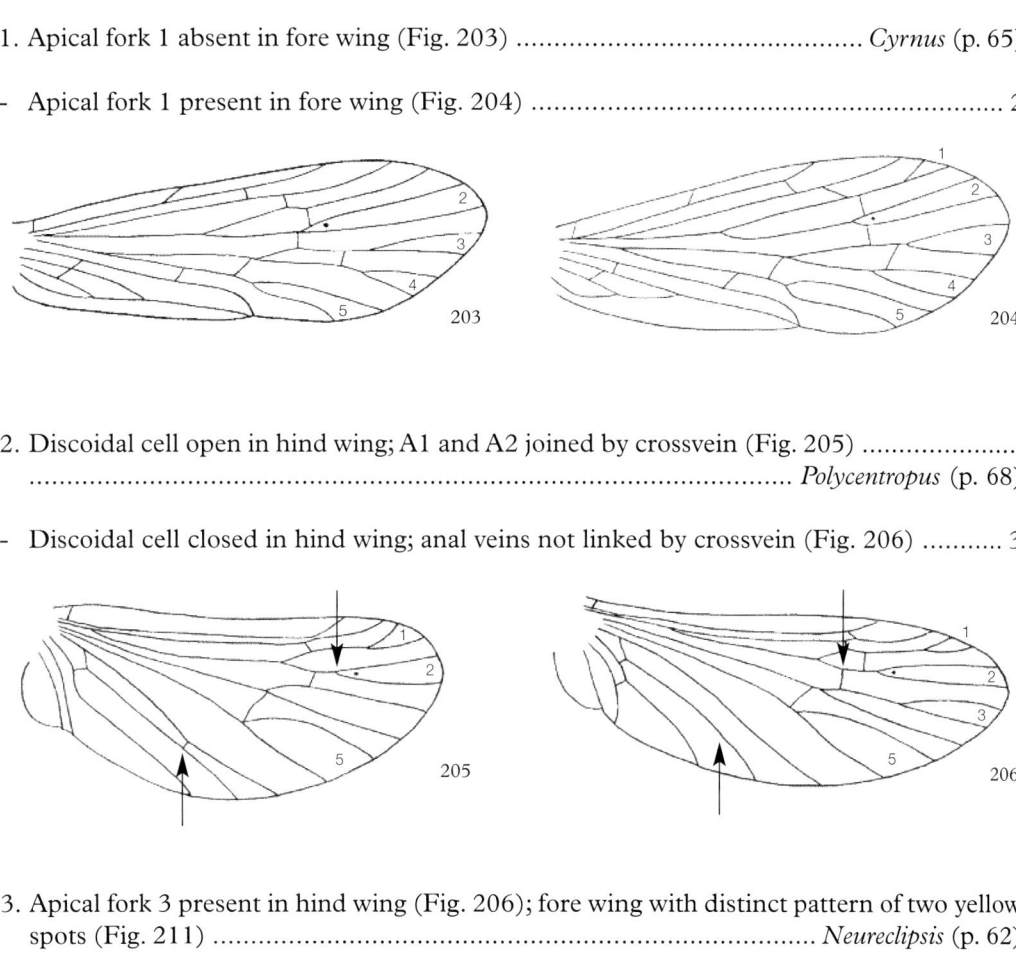

2. Discoidal cell open in hind wing; A1 and A2 joined by crossvein (Fig. 205) ......................
.................................................................................................... *Polycentropus* (p. 68)

- Discoidal cell closed in hind wing; anal veins not linked by crossvein (Fig. 206) ........... 3

3. Apical fork 3 present in hind wing (Fig. 206); fore wing with distinct pattern of two yellow spots (Fig. 211) .............................................................................. *Neureclipsis* (p. 62)

- Apical fork 3 absent in hind wing (Fig. 207) ................................................................. 4

4. Apical fork 1 present in hind wing (Fig. 207); fore wing usually more than 9 mm long ......
...................................................................................... *Plectrocnemia* (p. 70)

- Apical fork 1 absent in hind wing (Fig. 208); fore wing usually less than 9 mm long ..........
...................................................................................... *Holocentropus* (p. 63)

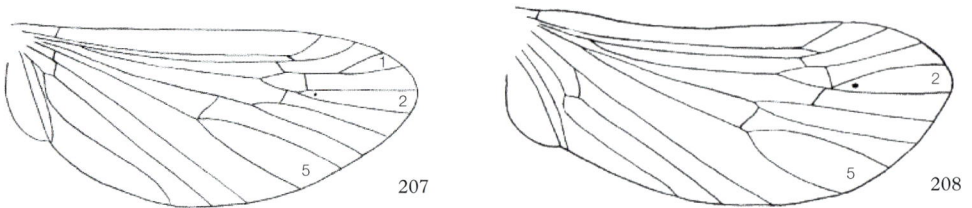

## Genus NEURECLIPSIS McLachlan, 1864

Fore wing with forks 1 to 5, hind wing with forks 1, 2, 3 and 5; discoidal cell closed in both wings (Fig. 209). *N. bimaculata* is the only European species.

### *Neureclipsis bimaculata* (Linnaeus, 1758)

Listed as *Neureclepsis* [sic] in Macan (1973). Fore wing length: ♂ 5-8 mm, ♀ 8-10 mm (Figs 209-210). The two yellow spots on the greyish brown fore wings are very conspicuous in life (Fig. 211). England except SW. but always local, Wales and mainland Scotland; present in Ireland; streams, rivers, and often lake outflows, where it can be abundant. Throughout Europe, except extreme south. Flight period: May-October. ♂ genitalia with characteristically narrow and pointed intermediate appendages (Fig. 212); ♀ genitalia (Fig. 213).

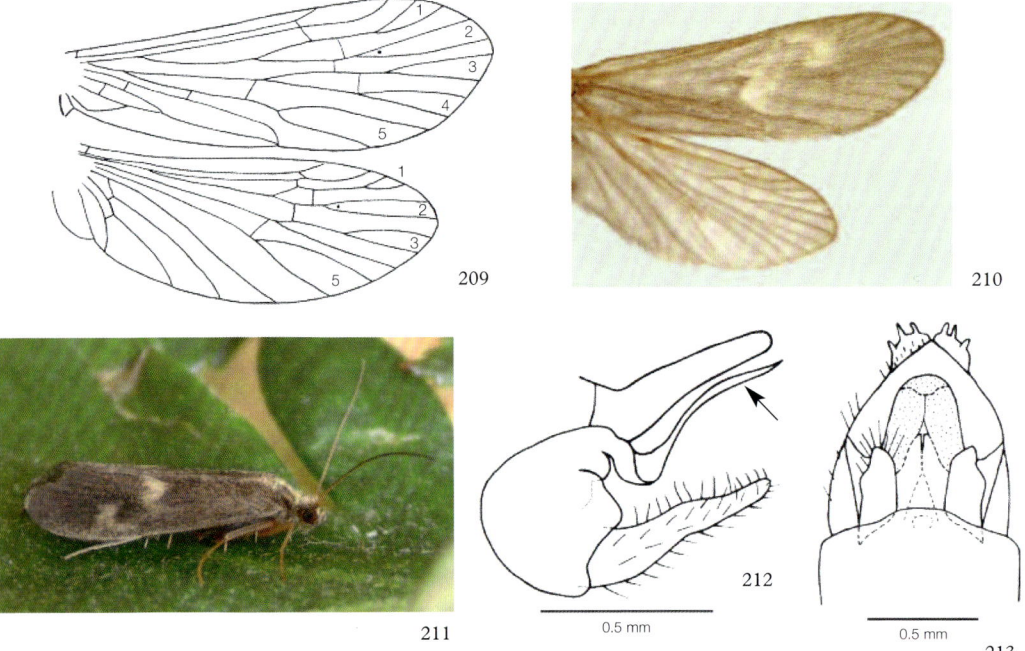

Figures 209-213. *Neureclipsis bimaculata*. 209 wing venation; 210 wing pattern; 211 live specimen; 212 male genitalia lateral, aedeagus omitted; 213 female genitalia ventral

# Genus HOLOCENTROPUS McLachlan, 1878

Fore wing with forks 1 to 5 present, hind wing with forks 2 and 5 only; discoidal cell closed in both wings (Fig. 218). Around five species in Europe of which three occur in Britain.

## *Holocentropus dubius* (Rambur, 1842)

Fore wing length: ♂ 6-8 mm, ♀ 6-9 mm (Figs 214, 215). Throughout Britain, but local; present in Ireland; still and slow-flowing water. Throughout much of Europe. Flight period: May-August. ♂ genitalia with characteristic long downcurved segment IX, clasper with distinct dorsal lobe (Fig. 216); ♀ genitalia with two pairs of ventral appendages (Fig. 217).

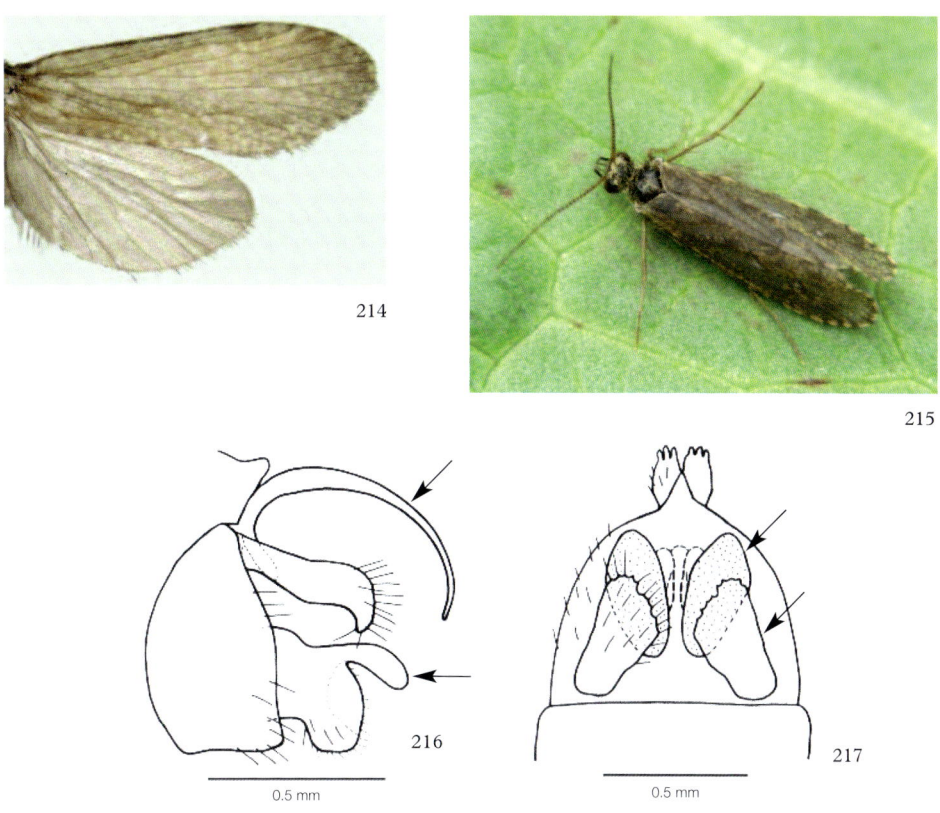

Figures 214-217. *Holocentropus dubius*. 214 wing pattern; 215 live specimen; 216 male genitalia lateral, aedeagus omitted; 217 female genitalia ventral

## *Holocentropus picicornis* (Stephens, 1836)

Fore wing length: ♂ 5-7 mm, ♀ 7-9 mm (Figs 218-220); the typical form has dark brown wings with yellow mottling (Fig. 219), but form *aurata* has unmarked yellowish brown wings (Fig. 220). Throughout Britain, present in Ireland; still and slow-flowing water. Throughout Europe, less common in south. Flight period: May-September. ♂ genitalia with simple broadly triangular claspers (Fig. 221); ♀ genitalia with narrow ventral appendages, broadest at base (Fig. 222).

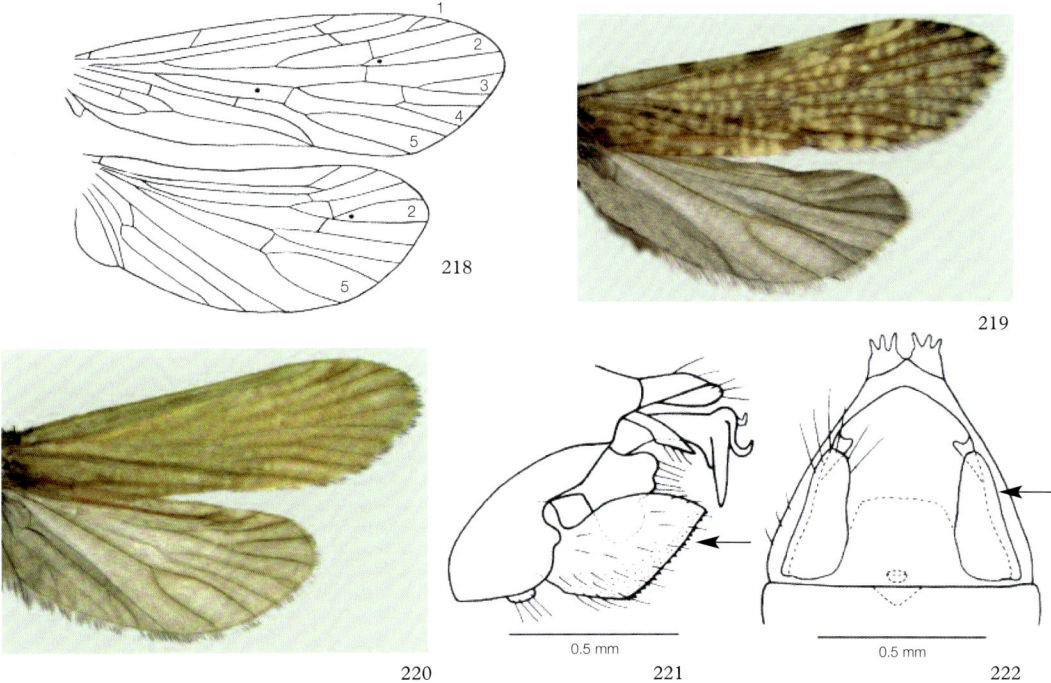

Figures 218-222. *Holocentropus picicornis*. 218 wing venation; 219 typical wing pattern; 220 form *aurata*; 221 male genitalia lateral, aedeagus omitted; 222 female genitalia ventral

## *Holocentropus stagnalis* (Albarda, 1874)

Fore wing length: ♂♀ 5-7 mm (Figs 223, 224). Very local, with scattered records from England and parts of Scotland, few records from Wales; no records from Ireland; still water. Mainly central Europe. Flight period: April-August. ♂ genitalia with dorsal process on claspers that curves sharply inwards (Figs 225, 226); ♀ genitalia with rather pointed ventral processes that are narrow at the base (Fig. 227).

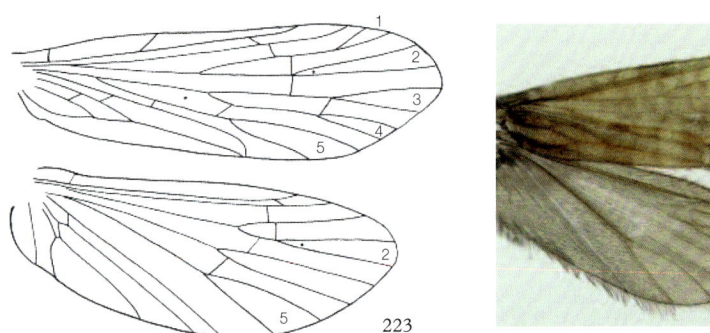

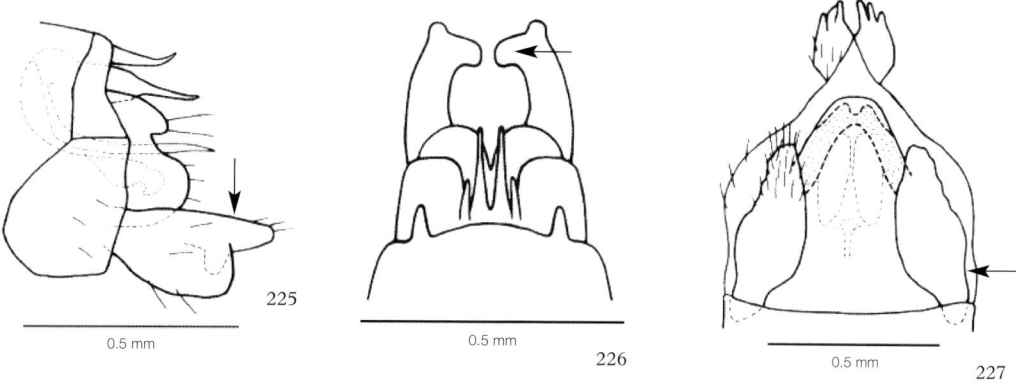

Figures 223-227. *Holocentropus stagnalis*. 223 wing venation; 224 wing pattern; 225 male genitalia lateral, aedeagus omitted; 226 male genitalia dorsal, aedeagus omitted; 227 female genitalia ventral

# Genus CYRNUS Stephens, 1836

Fore wing with forks 2 to 5 present, hind wing with forks 2 and 5 only; discoidal cell closed in fore wing, open in hind wing (Fig. 228). Non-genitalic characters for the separation of females in this genus were proposed by Robert & Neu (2002) but they do not seem to work consistently. Of about six species in Europe three are found in Britain.

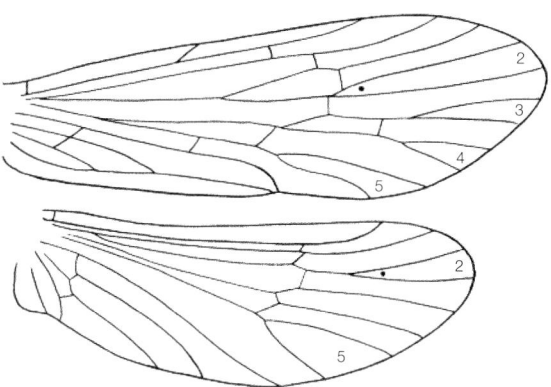

Figure 228. *Cyrnus flavidus*. Wing venation

## *Cyrnus flavidus* **McLachlan, 1864**

Fore wing length: ♂ 6-8 mm, ♀ 6-9 mm (Figs 228-230); a pale yellow species in life. Common throughout England except SW., Wales and Scotland; present in Ireland; still and slow-flowing water. Mainly central and northern Europe. Flight period: May-September. ♂ genitalia with only short appendages on tergite IX, claspers broad and rectangular in lateral view (Fig. 231); ♀ genitalia with broad parallel-sided ventral appendages (Fig. 232).

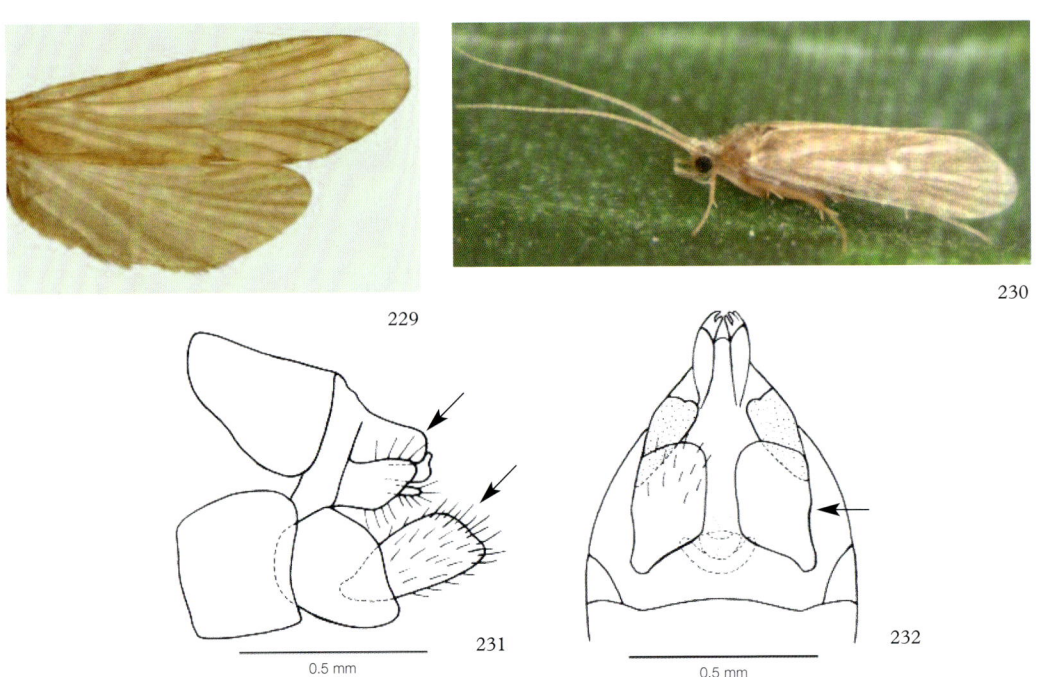

Figures 229-232. *Cyrnus flavidus*. 229 wing pattern; 230 live specimen; 231 male genitalia lateral, aedeagus omitted; 232 female genitalia ventral

## *Cyrnus trimaculatus* **(Curtis, 1834)**

Fore wing length: ♂ 5-7 mm, ♀ 5-9 mm (Figs 233-235); usually dark with yellow mottling but very variable. Common throughout Britain; present in Ireland; still and slow-flowing water. Throughout Europe. Flight period: May-September. ♂ genitalia with long appendages on segment IX (triangular in dorsal view), claspers slightly rounded apically (Fig. 236); ♀ genitalia with subtriangular ventral appendages (Fig. 237).

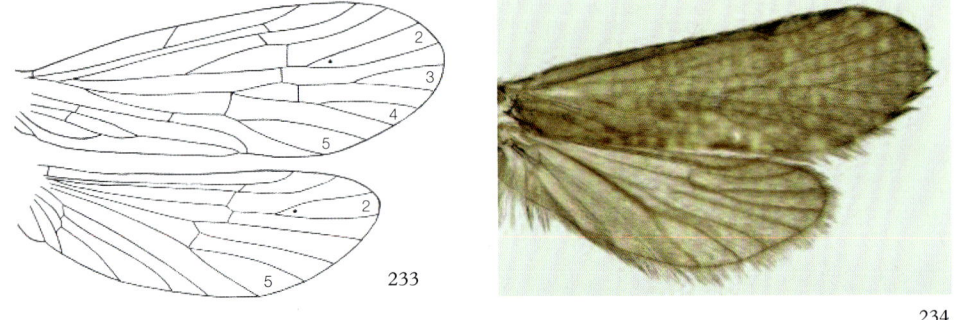

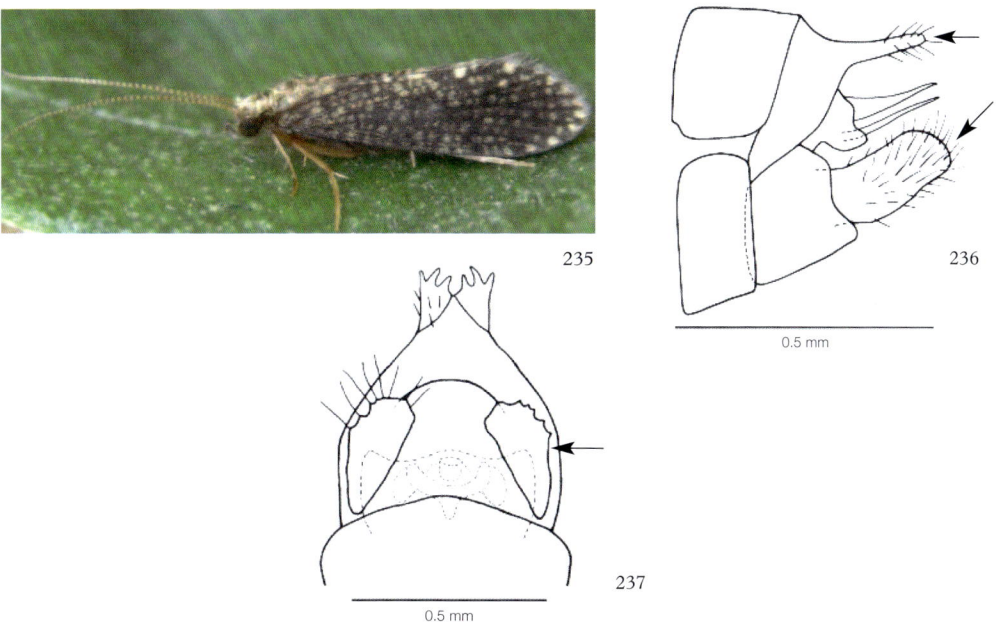

Figures 233-237. *Cyrnus trimaculatus.* 233 wing venation; 234 wing pattern; 235 live specimen; 236 male genitalia lateral, aedeagus omitted; 237 female genitalia ventral

## *Cyrnus insolutus* McLachlan, 1878

Not recorded as British until Kimmins (1942). Fore wing length: ♂♀ 5-6 mm (Fig. 238); less strongly marked than *C. trimaculatus.* Isolated sites in Cumbria and Hampshire, possibly also Wales; present in Ireland; lakes. Throughout much of Europe, less common in south. Flight period: July-August. ♂ genitalia with only short appendages on segment IX, claspers pointed in lateral view (Fig. 239); ♀ genitalia very similar to those of *C. trimaculatus* and no clear differences have yet been established (Fig. 240); however, most specimens of *C. insolutus* will be much paler and less strongly marked than *C. trimaculatus.*

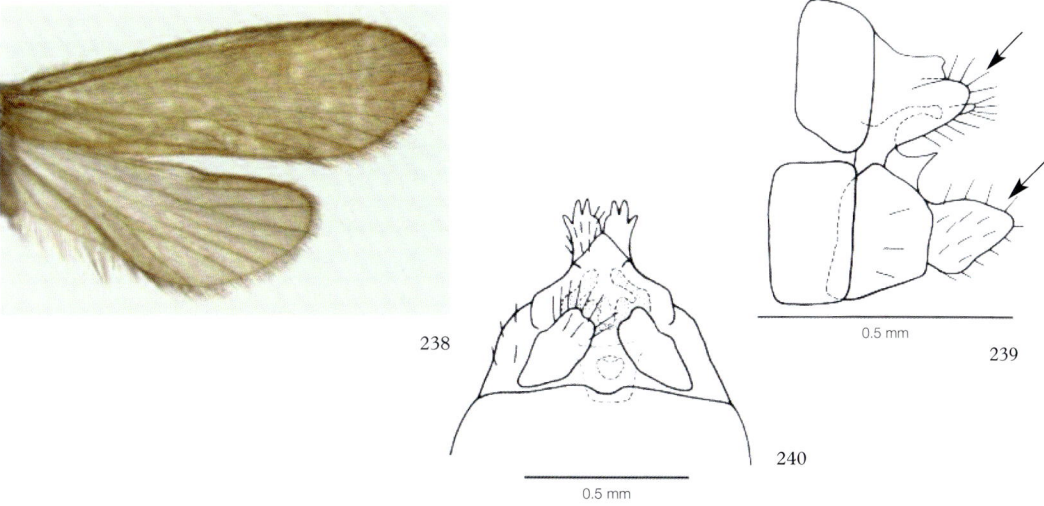

Figures 238-240. *Cyrnus insolutus.* 238 wing pattern; 239 male genitalia lateral, aedeagus omitted; 240 female genitalia ventral

# Genus POLYCENTROPUS Curtis, 1835

Fore wing with forks 1 to 5 present, hind wing with forks 1, 2 and 5; discoidal cell closed in fore wing, open in hind wing (Fig. 241). Note that the apparent differences in wing coloration between the three species are not always consistent. A large genus with around 30 species in Europe, but just three in Britain.

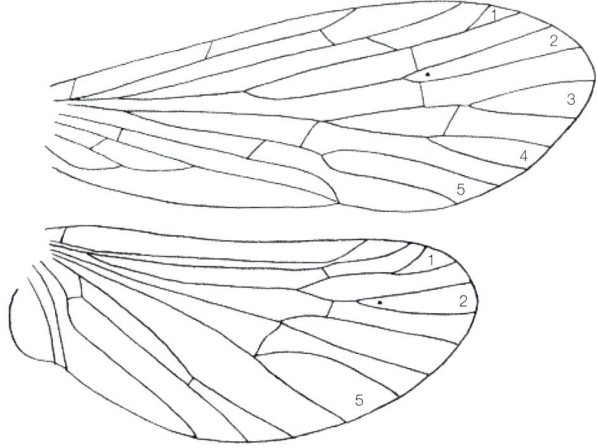

Figure 241. *Polycentropus flavomaculatus.* Wing venation

## *Polycentropus flavomaculatus* (Pictet, 1834)
= *multiguttatus* Curtis, 1835

Fore wing length: ♂ 6-9 mm, ♀ 9-12 mm (Figs 241-243). Common throughout Britain; present in Ireland; streams, rivers and stony lakes. Throughout Europe. Flight period: May-September, possibly bivoltine. ♂ genitalia with intermediate appendages hooked sharply downwards in lateral view (Fig. 244) and divergent in dorsal view (Fig. 245); ♀ genitalia with flattened apex to subgenital plate, ventral appendages short (Fig. 246).

243

242

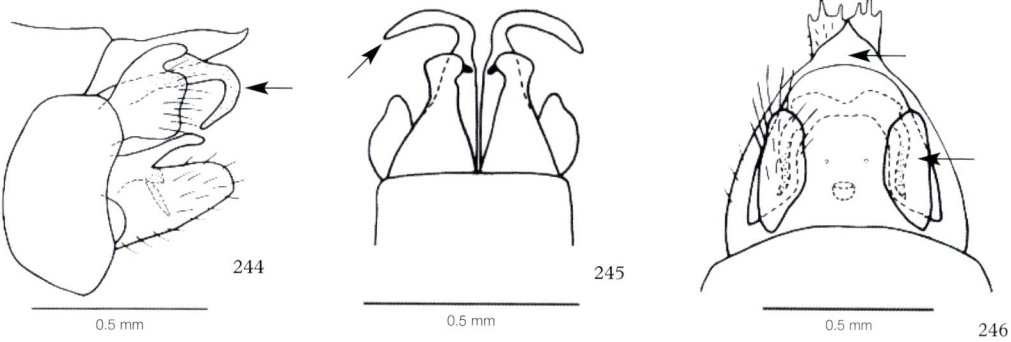

Figures 242-246. *Polycentropus flavomaculatus*. 242 wing pattern; 243 live specimen;
244 male genitalia lateral, aedeagus omitted; 245 male genitalia dorsal, aedeagus omitted; 246 female genitalia ventral

## *Polycentropus irroratus* (Curtis, 1835)
= *multiguttatus*; misidentified by some authors

Listed as *P. multiguttatus* in Mosely (1939). Fore wing length: ♂ 7-9 mm, ♀ 9-10 mm (Fig. 247). Throughout Britain but local; present in Ireland; streams, rivers and lakes. Scattered throughout much of Europe. Flight period: May-September. ♂ genitalia with intermediate appendages looped downwards, superior appendages large and triangular (Fig. 248); ♀ genitalia with rounded apex to subgenital plate, ventral appendages short (Fig. 249).

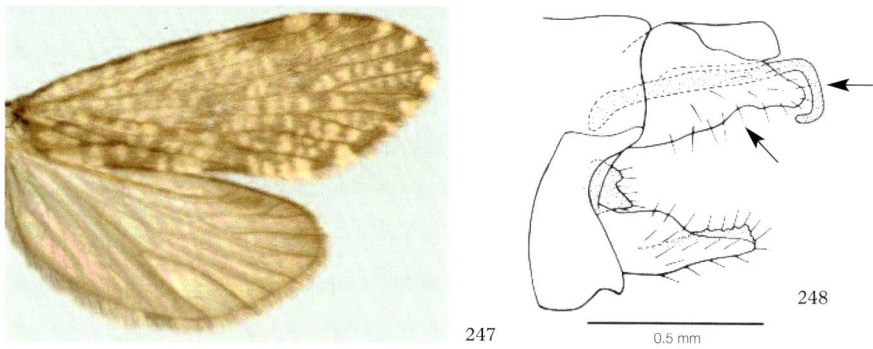

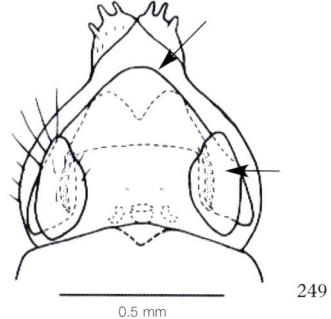

Figures 247-249. *Polycentropus irroratus*. 247 wing pattern;
248 male genitalia lateral, aedeagus omitted; 249 female genitalia ventral

## *Polycentropus kingi* McLachlan, 1881

Fore wing length: ♂ 6-8 mm, ♀ 8-10 mm (Fig. 250). Can be moderately common in Scotland, Wales, N. and SW. England; present in Ireland; stony rivers and streams. Mainly south-western Europe. Flight period: June-September, and usually on the wing for a very short time in any particular locality. Possibly over-recorded as a larva because of identification problems; it is always less abundant than *P. flavomaculatus* at any site. ♂ genitalia with intermediate appendages bent upwards in lateral view (Fig. 251); ♀ genitalia with rounded or slightly bifurcate apex to subgenital plate, ventral appendages much longer than in the other two species (Fig. 252).

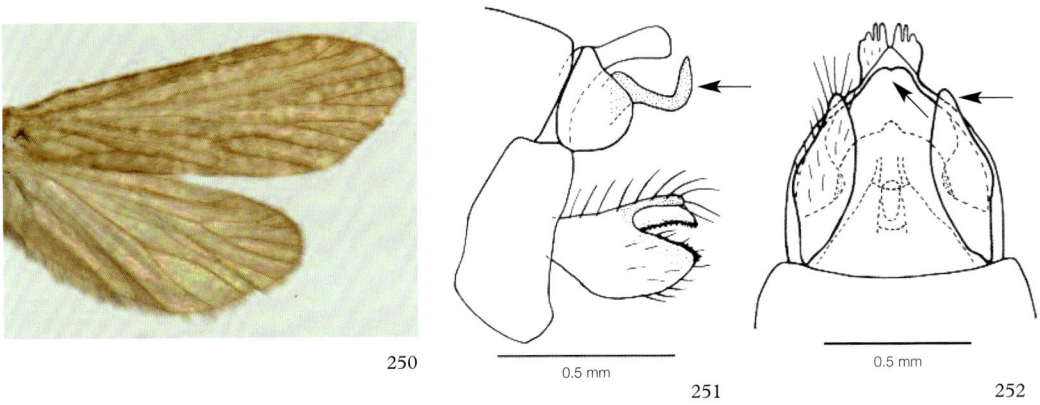

250    0.5 mm    0.5 mm

251    252

Figures 250-252. *Polycentropus kingi.* 250 wing pattern;
251 male genitalia lateral, aedeagus omitted; 252 female genitalia ventral

# Genus PLECTROCNEMIA Stephens, 1836

Fore wing with forks 1 to 5 present, hind wing with forks 1, 2 and 5; discoidal cell closed in both wings (Fig. 253). Around 15 European species, with just three in Britain. A key to the European species was given by Roy *et al.* (1980).

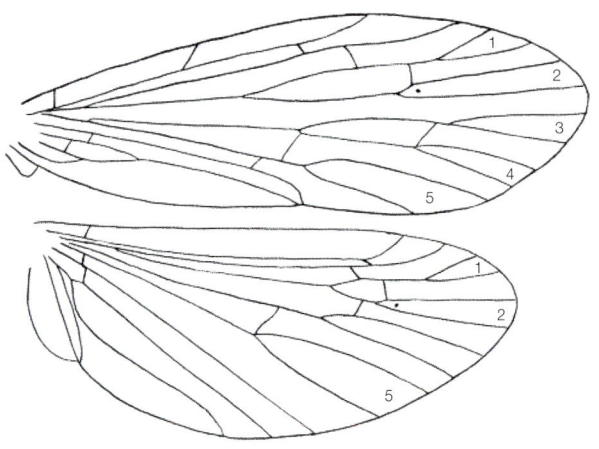

Figure 253. *Plectrocnemia conspersa.* Wing venation

## *Plectrocnemia conspersa* (Curtis, 1834)

Fore wing length: ♂ 9-14 mm, ♀ 9-15 mm (Figs 253-254). Common throughout Britain; present in Ireland; trickles to small streams, small stony lakes, permanently flowing marshes. Throughout Europe. Flight period: May-September. ♂ genitalia with very narrow and pointed ventral branch to clasper (Fig. 255); ♀ genitalia with narrow subgenital plate that is much longer than ventral appendages (Fig. 256).

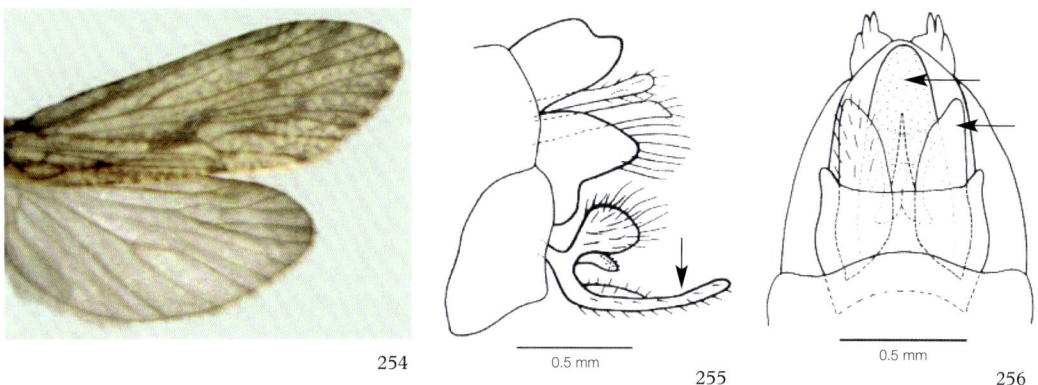

Figures 254-256. *Plectrocnemia conspersa*. 254 wing pattern;
255 male genitalia lateral, aedeagus omitted; 256 female genitalia ventral

## *Plectrocnemia brevis* McLachlan, 1871

Fore wing length: ♂♀ 9-10 mm (Fig. 257); usually paler than the other two species. Very local in Devon, Wales, and N. England; no records from Ireland; streams and permanent trickles that are calcareous and usually depositing. Mainly southern and central Europe. Flight period: May-July. ♂ genitalia with short and stout ventral branch to clasper (Fig. 258); ♀ genitalia with broad subgenital plate that is only slightly longer than ventral appendages (Fig. 259).

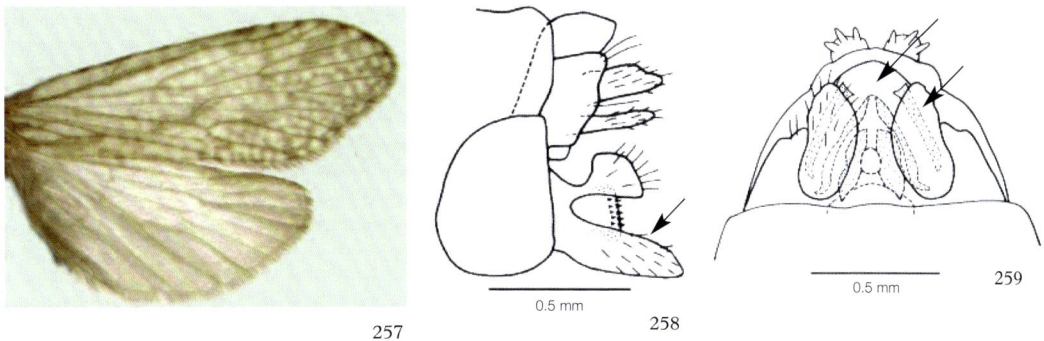

Figures 257-259. *Plectrocnemia brevis*. 257 wing pattern;
258 male genitalia lateral, aedeagus omitted; 259 female genitalia ventral

### *Plectrocnemia geniculata* **McLachlan, 1871**

Fore wing length: ♂♀ 9-13 mm (Fig. 260). Common in SW. and N. England, Wales and the bordering English counties, Scotland, with a few isolated records for other parts of England; present in Ireland; streams and trickles. Mainly southern and central Europe. Flight period: April-September, possibly bivoltine. ♂ genitalia with elongate apex but not divided into dorsal and ventral branches (Fig. 261); ♀ genitalia with triangular subgenital plate and very small ventral appendages (Fig. 262).

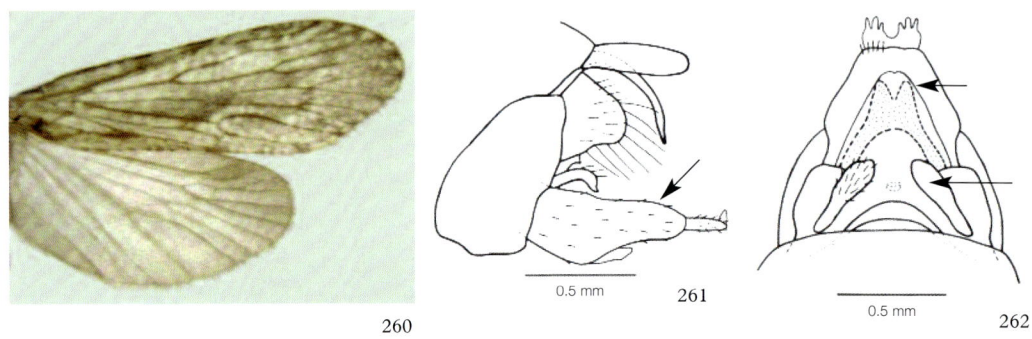

0.5 mm    261

0.5 mm    262

260

Figures 260-262. *Plectrocnemia geniculata*. 260 wing pattern;
261 male genitalia lateral, aedeagus omitted; 262 female genitalia ventral

# Family PSYCHOMYIIDAE (3 genera, 12 species)

Wing venation given under each genus below. Spur formula 2.4.4. Ocelli absent. The family name was often mis-spelled as Psychomyidae by earlier authors.

## Key to genera of Psychomyiidae

1. Apical fork 3 in hind wing sessile (Fig. 263) ........................................... *Tinodes* (p. 77)

- Apical fork 3 in hind wing stalked (Fig. 264) ............................................................. 2

263    264

2. Wings black (in life) ................................................................................. *Lype* (p. 73)

- Wings brown (in life) ......................................................................... *Psychomyia* (p. 75)

# Genus LYPE McLachlan, 1878

Fore wing with forks 2 to 5 present, hind wing with forks 2, 3 and 5; discoidal cell normally closed in both wings, though the crossvein is often hard to see (Fig. 265). These are the angler's Micro Black Sedges. Two of the three European species occur in Britain.

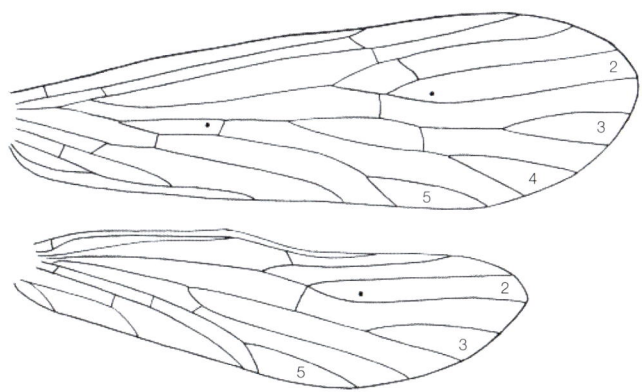

Figure 265. *Lype phaeopa.* Wing venation

## *Lype phaeopa* (Stephens, 1836)

Fore wing length: ♂♀ 4-6 mm (Figs 265, 266). Note the apparent differences in venation between this species and *L. reducta*, in that the latter is reported to have the discoidal cell closed in the hind wing. Mosely (1939) shows this cell as open in *L. phaeopa* (cf. Figs 265 and 269) but the crossvein is extremely faint and this distinction should not be considered as reliable. Common throughout Britain; present in Ireland; mainly rivers and larger streams. Throughout Europe. Flight period: (April) May-September. ♂ genitalia with dorsal process sharply bent in lateral view (Fig. 267); ♀ genitalia with slight indentation in dorsal margin of segment IX in lateral view (Fig. 268).

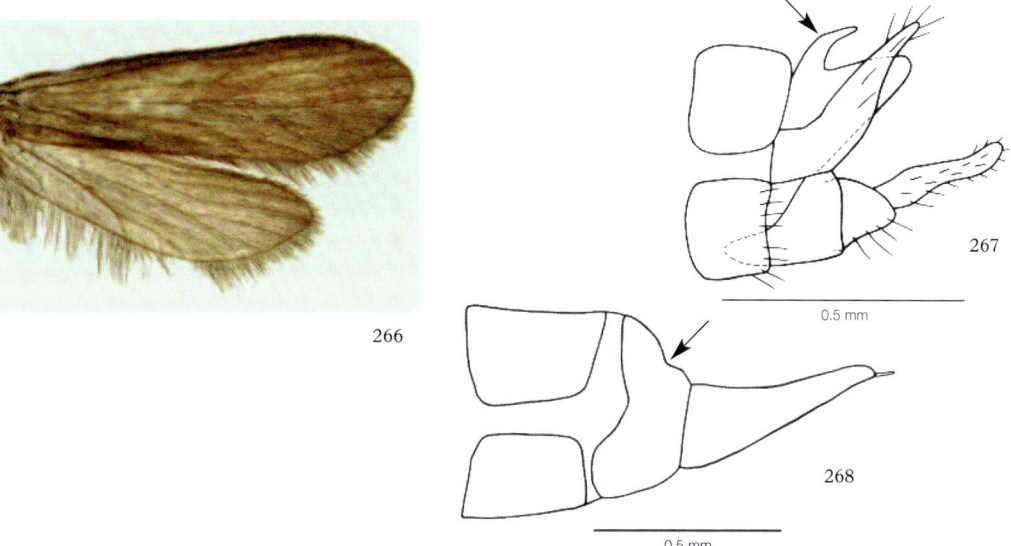

Figures 266-268. *Lype phaeopa.* 266 wing pattern; 267 male genitalia lateral; 268 female genitalia lateral

## *Lype reducta* (Hagen, 1868)

Fore wing length: ♂♀ 4-6 mm (Figs 269, 270). See the notes on venation under *L. phaeopa* (above). Records from England, Wales and southern Scotland, local but status not clear because of uncertainty over some larval records; present in Ireland; mainly small streams. Throughout Europe. Flight period: May-August. ♂ genitalia with dorsal process evenly curved in lateral view (Fig. 271); ♀ genitalia with evenly curved dorsal margin of segment IX in lateral view (Fig. 272).

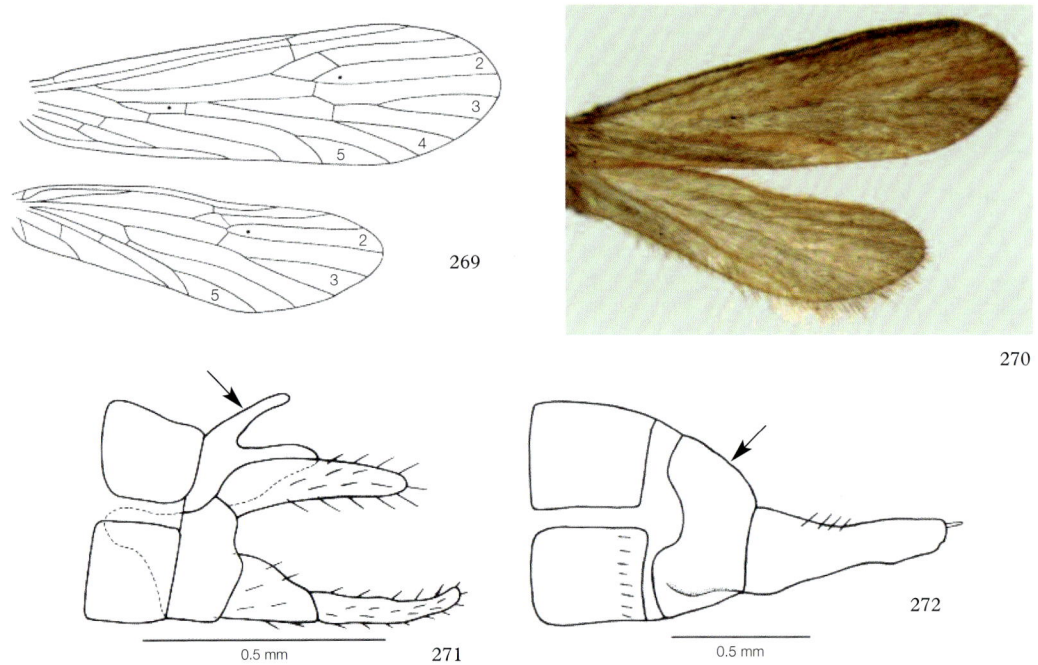

Figures 269-272. *Lype reducta*. 269 wing venation; 270 wing pattern; 271 male genitalia lateral; 272 female genitalia lateral

# Genus PSYCHOMYIA Latreille, 1829
# = METALYPE Klapálek, 1898

Fore wing with forks 2 to 5 present, hind wing with forks 2, 3 and 5; discoidal cell closed in fore wing, open in hind wing (Fig. 273). Two of the six European species occur in Britain.

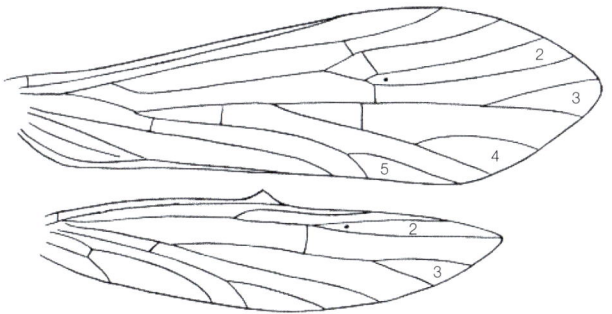

Figure 273. *Psychomyia pusilla.* Wing venation

## *Psychomyia pusilla* (Fabricius, 1781)

This is the angler's Small Yellow Sedge. Fore wing length: ♂♀ 3-5 mm (Figs 273, 274). Common throughout England, Wales and mainland Scotland; present in Ireland; rivers and large streams. Throughout Europe. Flight period: May-September. ♂ genitalia with sinuous superior appendages, directed upwards (Fig. 275); ♀ genitalia with upwardly curved dorsal margin of segment IX in lateral view (Fig. 276), segment IX short in ventral view (Fig. 277).

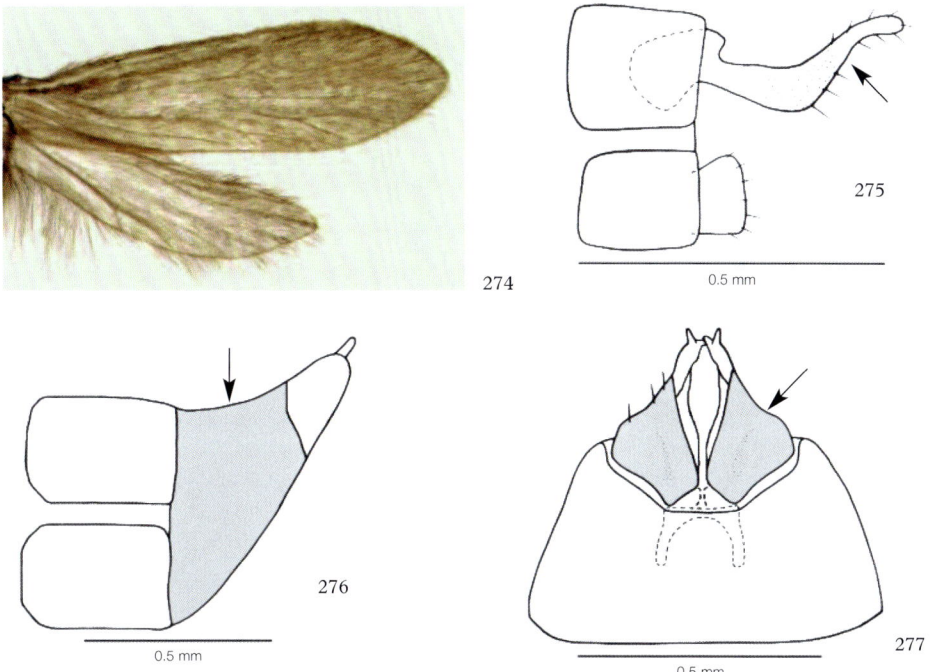

Figures 274-277. *Psychomyia pusilla.* 274 wing pattern; 275 male genitalia lateral; 276 female genitalia lateral; 277 female genitalia ventral, segment IX shaded

## *Psychomyia fragilis* (Pictet, 1834)

Listed as *Metalype fragilis* in all previous British works. *Metalype* can no longer be considered as distinct from *Psychomyia*, and it was formally synonymised by Malicky (1995). Fore wing length: ♂♀ 4-5 mm (Figs 278, 279). Very local, with records from Somerset, Gloucestershire, Hampshire, Derbyshire and NW. England; present in Ireland; calcareous streams, rivers and lakes. Mainly southern and western Europe. Flight period: June-September. ♂ genitalia with relatively straight superior appendages that terminate in an apical inward-pointing tooth (Fig. 280); ♀ genitalia with flat dorsal margin of segment IX in lateral view (Fig. 281), segment IX long in ventral view (Fig. 282).

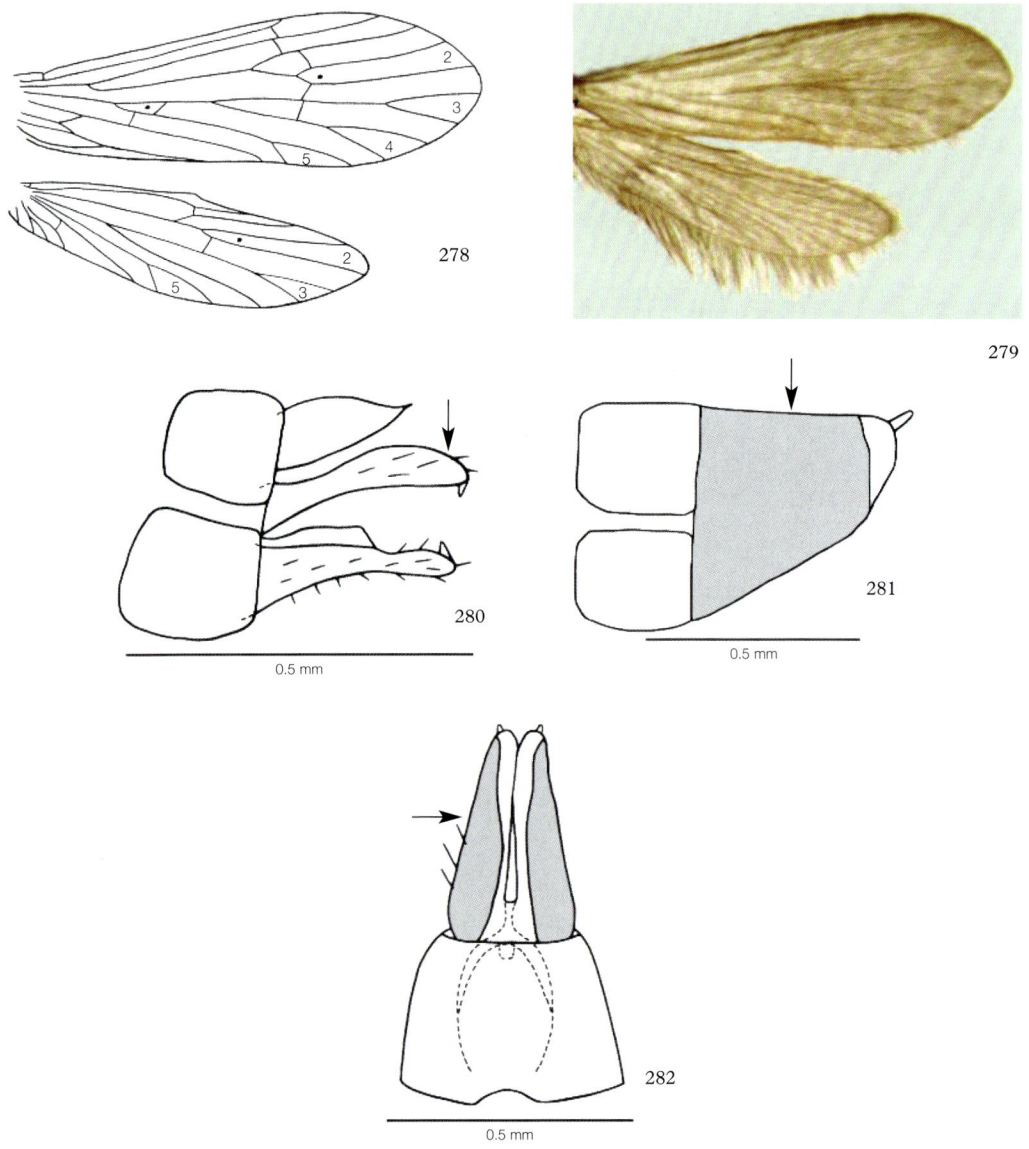

Figures 278-282. *Psychomyia fragilis*. 278 wing venation; 279 wing pattern; 280 male genitalia lateral; 281 female genitalia lateral; 282 female genitalia ventral, segment IX shaded

# Genus TINODES Leach, 1815

Fore wing with forks 2 to 5 present, hind wing with forks 2, 3 and 5; discoidal cell closed in fore wing, open in hind wing (Fig. 283). A large genus, with around 60 species in Europe, but just eight in the British Isles, of which one is found only in Ireland. The complex male genitalia are not described in the text, but the critical characters are indicated by arrows on the accompanying figures. The illustrations of female genitalia are all from Fisher (1977).

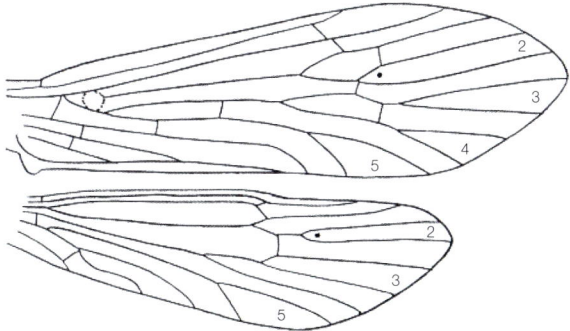

Figures 283. *Tinodes waeneri.* Wing venation

## *Tinodes waeneri* (**Linnaeus, 1758**)

This is the angler's Small Red Sedge. Fore wing length: ♂ 5-8 mm, ♀ 5-9 mm (Figs 283-285); in life the fore wings are reddish brown with darker veins. Common throughout Britain; present in Ireland; streams, rivers and lakes. Throughout most of Europe. Flight period: May-September, possibly bivoltine in the south. ♂ genitalia as in Fig. 286; ♀ genitalia easily distinguished from all others in the genus by the short 'ovipositor' formed from segment IX (Figs 287, 288). In the other species of *Tinodes* this segment is elongate and upcurved.

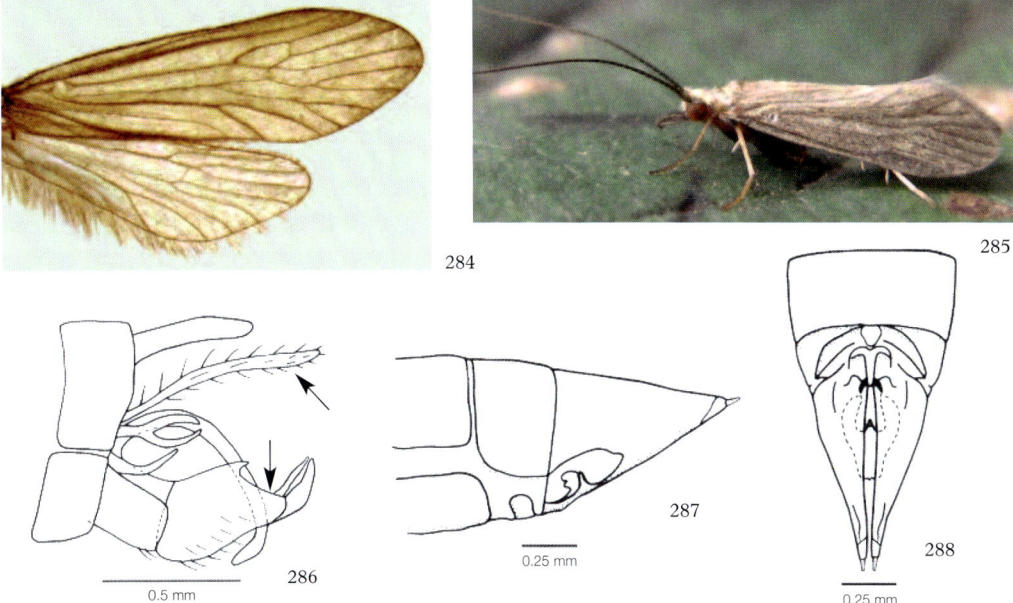

Figures 284-288. *Tinodes waeneri.* 284 wing pattern; 285 live specimen; 286 male genitalia lateral, aedeagus omitted; 287 female genitalia lateral; 288 female genitalia ventral

## *Tinodes rostocki* **McLachlan, 1878**

Fore wing length: ♂ 4-6 mm ♀ 6-8 mm; wings black in life. Very local in SE. Wales, Wyre Forest, Surrey, Essex and Derbyshire; no records from Ireland; rocky streams in woodland. Mainly southern and central Europe. Flight period: May-July. ♂ genitalia as in Fig. 289; ♀ genitalia with pair of 'pits' separated by a triangular membranous region (Figs 290, 291).

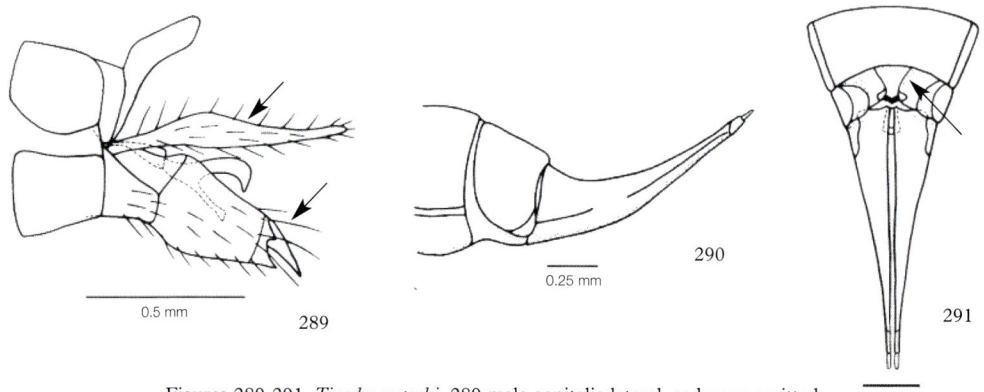

Figures 289-291. *Tinodes rostocki.* 289 male genitalia lateral, aedeagus omitted; 290 female genitalia lateral; 291 female genitalia ventral

## *Tinodes dives* **(Pictet, 1834)**

Fore wing length: ♂ ♀ 5-6 mm (Figs 292, 293); wings black with large pale yellowish patch near apex. Local in S. and N. Wales, Derbyshire and N. England, with scattered records elsewhere from England and central Scotland; present in Ireland; calcareous headwater streams in limestone moorland. Mainly southern and central Europe. Flight period: May-August. Unlike other members of *Tinodes* this is a day-flying species and the pale wing markings are probably important for visual recognition. ♂ genitalia as in Fig. 294; ♀ genitalia with triangular segment VIII in lateral view (Figs 295, 296).

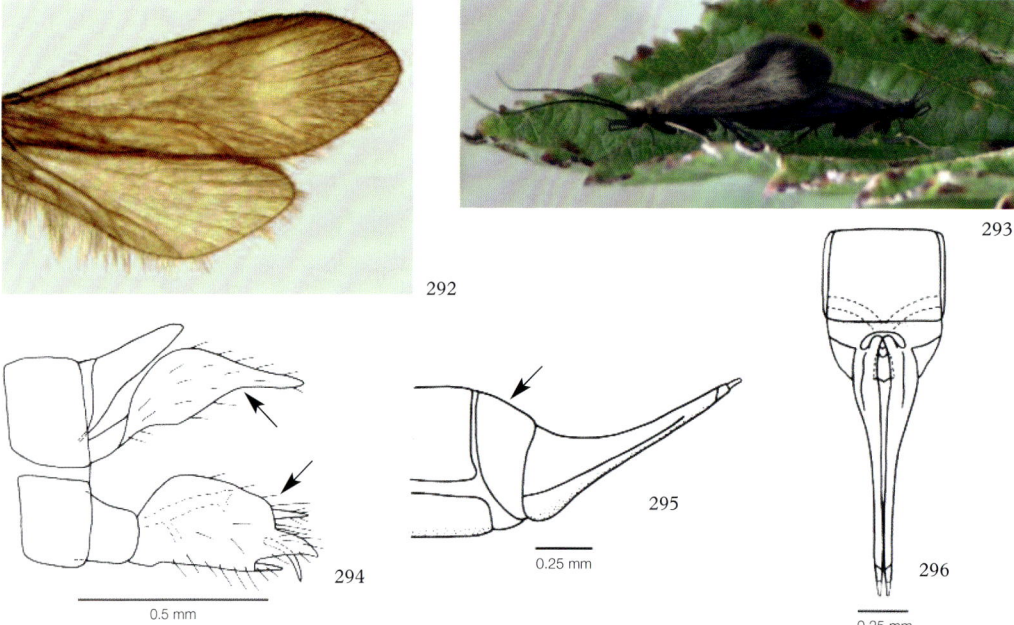

Figures 292-296. *Tinodes dives.* 292 wing pattern; 293 mating pair; 294 male genitalia lateral, aedeagus omitted; 295 female genitalia lateral; 296 female genitalia ventral

## *Tinodes maculicornis* (Pictet, 1834)

Fore wing length: ♂♀ 5-6 mm; wings pale greyish yellow in life. Ireland only (O'Connor & Wise, 1980). Mainly south-western Europe. ♂ genitalia as in Fig. 297; ♀ genitalia with well-developed sternite VIII clearly visible in lateral view (Fig. 298) and with a sclerotised sinuous posterior margin in ventral view (Fig. 299).

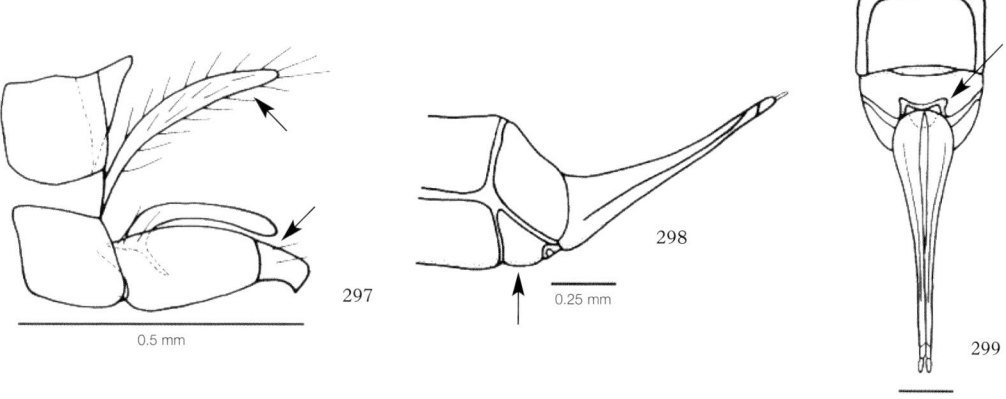

Figures 297-299. *Tinodes maculicornis.* 297 male genitalia lateral, aedeagus omitted; 298 female genitalia lateral; 299 female genitalia ventral

## *Tinodes unicolor* (Pictet, 1834)

Fore wing length: ♂ 4-8 mm, ♀ 4-6 mm; wings pale greyish yellow in life. Very local in England and Wales; present in Ireland; calcareous streams. Mainly southern and central Europe. Flight period: June-September. ♂ genitalia as in Fig. 300; ♀ genitalia similar to *T. maculicornis* but posterior margin of segment VIII not sclerotised (Figs 301, 302).

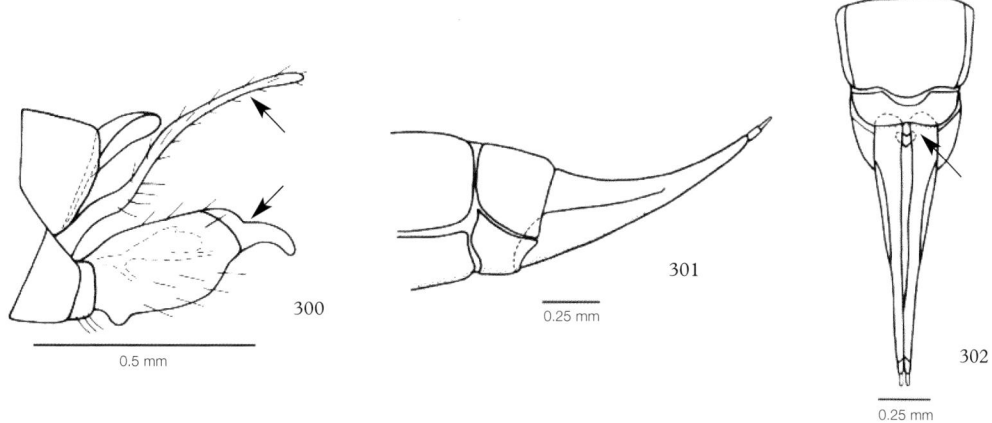

Figures 300-302. *Tinodes unicolor.* 300 male genitalia lateral, aedeagus omitted; 301 female genitalia lateral; 302 female genitalia ventral

## *Tinodes maclachlani* **Kimmins, 1966**
= *pusillus* Walker, 1852

Listed as *T. aureola* in Mosely (1939). Fore wing length: ♂♀ 6-7 mm (Fig. 303); wings pale greyish yellow in life. Local in SW. England, north midlands, N. England, Wales and Scotland; present in Ireland; trickles on rock faces or small stony streams. Mainly southern and central Europe. Flight period: May-August. ♂ genitalia as in Fig. 304; ♀ genitalia with a pair of triangular ventral lobes on sternum VIII that have strongly darkened margins (Figs 305, 306).

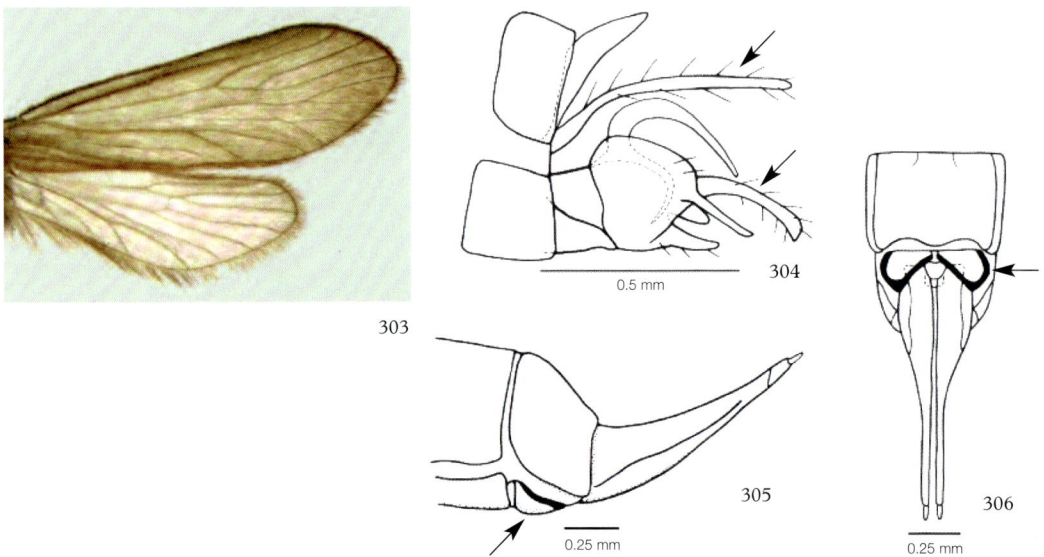

Figures 303-306. *Tinodes maclachlani.* 303 wing pattern; 304 male genitalia lateral, aedeagus omitted; 305 female genitalia lateral; 306 female genitalia ventral

## *Tinodes assimilis* **McLachlan, 1865**

Fore wing length: ♂♀ 6-7 mm; wings pale greyish yellow in life. Local in SW., S. and NW. England, Wales and western Scotland, with a few other isolated records; no records from Ireland; trickles on vertical rock faces. Mainly southern and western Europe. Flight period: May-August. ♂ genitalia as in Fig. 307; ♀ genitalia with triangular ventral lobes on sternum VIII as in *T. maclachlani* but without dark margins (Figs 308, 309).

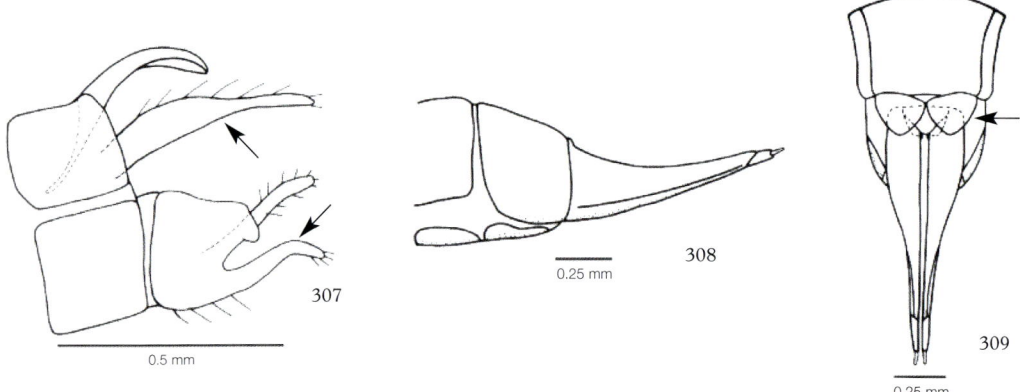

Figures 307-309. *Tinodes assimilis.* 307 male genitalia lateral, aedeagus omitted; 308 female genitalia lateral; 309 female genitalia ventral

## *Tinodes pallidulus* **McLachlan, 1878**

Not recognised as British until Kimmins (1949). Listed as *T. pallidula* in Macan (1973). Fore wing length: ♂♀ 5-6 mm; wings pale greyish yellow in life. Very local, with few records from Surrey, Kent. Leicestershire, Wyre Forest and Wales; no records from Ireland; small stony streams. Mainly central Europe. Flight period: July-August. ♂ genitalia as in Fig. 310; ♀ genitalia with a pair of lateral 'pits' that are visible in lateral view (Figs 311, 312).

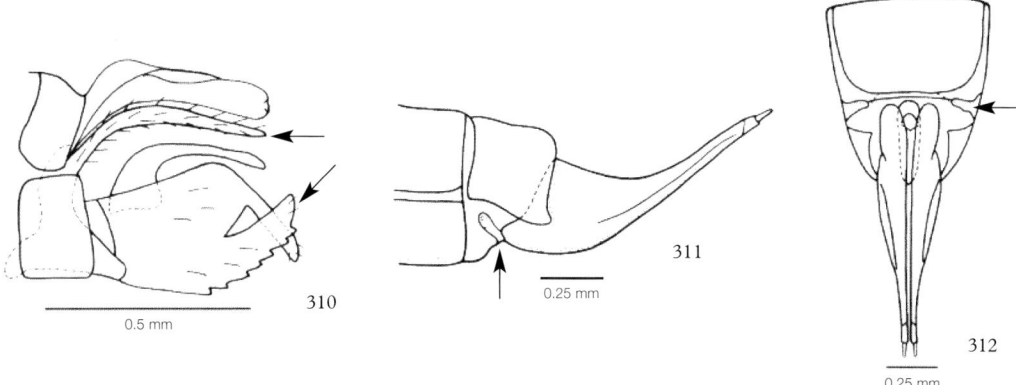

Figures 310-312. *Tinodes pallidulus*. 310 male genitalia lateral, aedeagus omitted; 311 female genitalia lateral; 312 female genitalia ventral

# Family HYDROPSYCHIDAE (3 genera, 11 species)

Fore wing with forks 1 to 5 present, hind wing with forks 1, 2, 3 and 5; discoidal cell closed in both wings. Spur formula 2.4.4. Ocelli absent.

## Key to genera of Hydropsychidae

1. Median cell in hind wing closed (Fig. 313); antennae with spiral ridge ............................
.................................................................................................... *Hydropsyche* (p. 84)

- Median cell in hind wing open (Fig. 314); antennae without spiral ridge ..................... 2

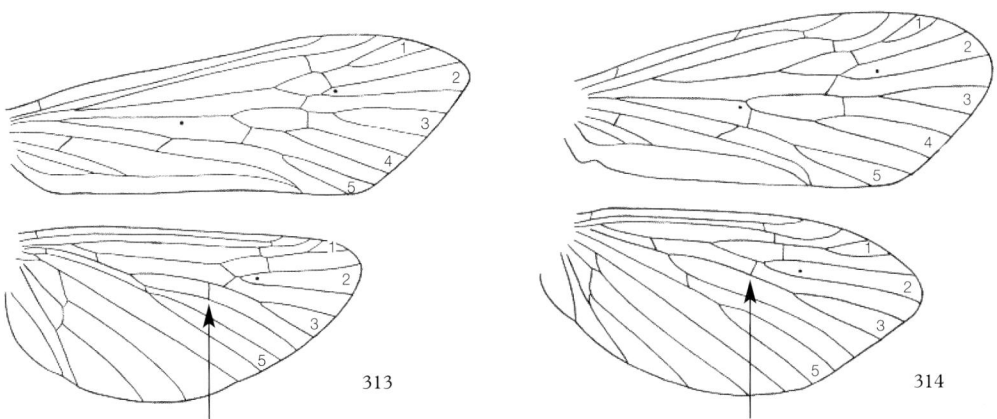

2. Both wings broad and rounded (Fig. 314), abdomen with long lateral filaments ...............
........................................................................................ *Diplectrona* (p. 82)

- Wings narrow (Fig. 315); abdomen without lateral filaments ........ *Cheumatopsyche* (p. 83)

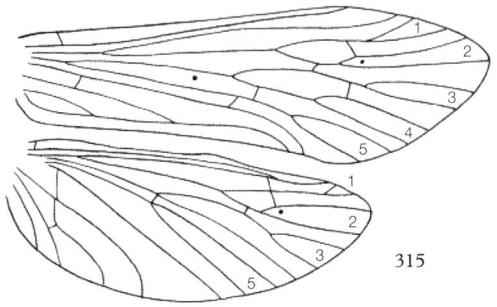

315

# Genus DIPLECTRONA Westwood, 1840

Just one of the 10 European species is found in Britain.

## *Diplectrona felix* **McLachlan, 1878**

Fore wing length: ♂ 6-7 mm, ♀ 7-9 mm (Figs 316-318): distinguished from all other members of the family by the broad rounded wings; fore wings brown with yellow spots. SW. England, Wales and the bordering English counties, N. England and mainland Scotland, with scattered records from Kent; present in Ireland; small cool streams. Southern and central Europe. Flight period: June-September. ♂ genitalia as in Fig. 319; ♀ genitalia as in Fig. 320.

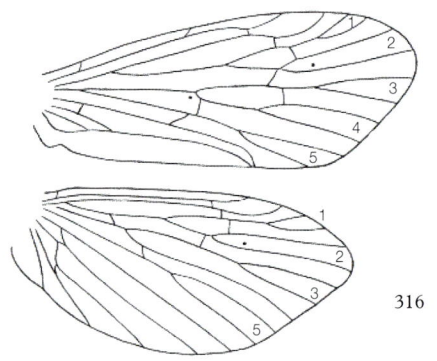

316

317

318

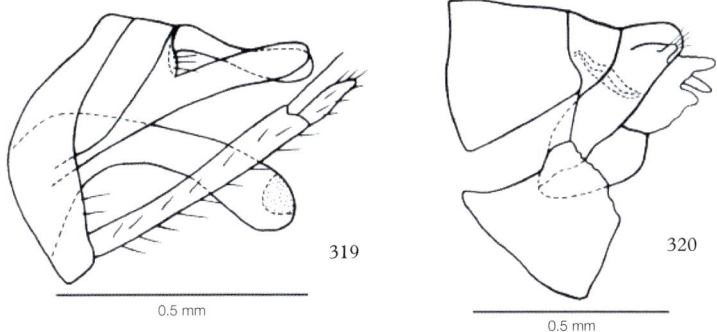

Figures 316-320. *Diplectrona felix*. 316 wing venation; 317 wing pattern; 318 live specimen; 319 male genitalia lateral; 320 female genitalia lateral

# Genus CHEUMATOPSYCHE Wallengren, 1891

Around eight European species, with just one in Britain.

## *Cheumatopsyche lepida* (Pictet, 1834)

Fore wing length: ♂ 5-7 mm, ♀ 6-8 mm (Figs 321, 322); fore wing superficially resembling *Hydropsyche* but with streaks of colour rather than spots, and there are no oblique ridges on the antennae. England (except East Anglia), Wales and mainland Scotland but always local; present in Ireland; rivers. Throughout Europe. Flight period: June-September. ♂ genitalia like those of *Hydropsyche* but with apex of clasper ending in a twisted hook (Fig. 323); ♀ genitalia with simple comma-shaped clasper receptacle (Fig. 324).

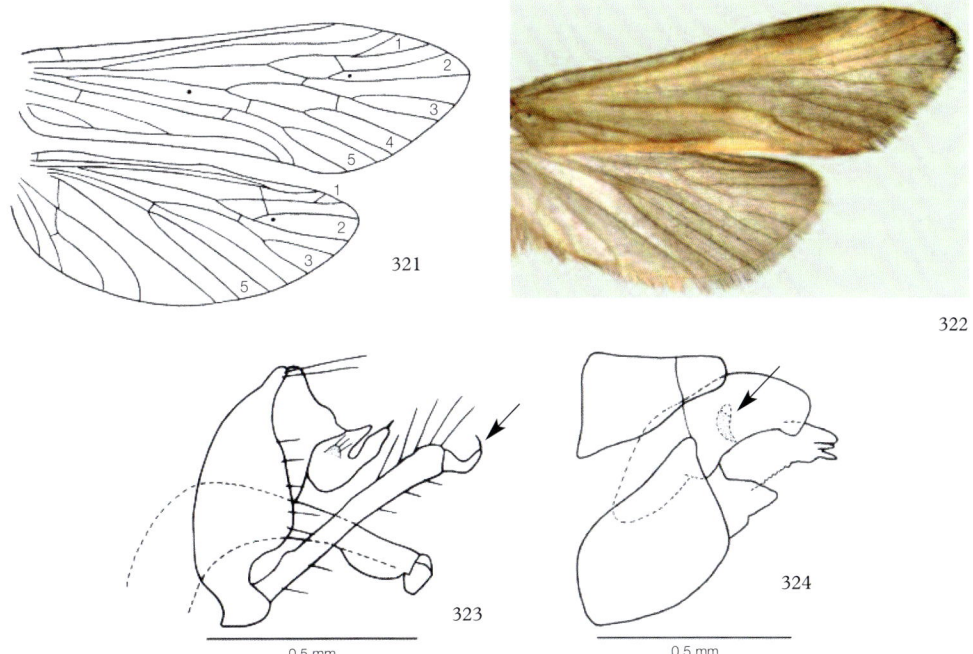

Figures 321-324. *Cheumatopsyche lepida*. 321 wing venation; 322 wing pattern; 323 male genitalia lateral; 324 female genitalia lateral

# Genus HYDROPSYCHE Pictet, 1834

These are the angler's Grey Flags or Marbled Sedges. This is a large and taxonomically complex genus with around 60 species in Europe, and nine in Britain. Many previous records were misidentified, and works such as Mosely (1939) and Macan (1973) cannot be used because of earlier confusion of species. Hildrew & Morgan's (1974) key is useful for males of *Hydropsyche* although the nomenclature is no longer up to date; see also Badcock (1977, 1978). Useful recent work in Europe includes Neu & Tobias (2004); for females see also Rojas-Camousseight *et al.* (1991).

Although loose species groups have been erected on the basis of male genitalia, there is a suggestion that female genitalia do not necessarily provide corroborative evidence for such groups (Mey, 2007). Examination of the genitalia in *Hydropsyche*, especially the females, will nearly always depend on clearing the abdomen in KOH; only very experienced workers can identify females simply preserved in alcohol, for example, because the internal structure of the clasper receptacles is often the main diagnostic character.

See Fig. 325 for typical wing venation in this genus. In the male genitalia the tenth segment has a median keel, known as the dorsal carina, and in some species this segment also bears a pair of setose processes. In the female the main diagnostic characters are the so-called dorsal and ventral lobes situated laterally on segment IX (shaded in the accompanying figures) and especially the clasper receptacle, a small but complex invagination situated just above the dorsal lobe in lateral view.

## *Hydropsyche instabilis* (Curtis, 1834)

Sometimes misidentified as *H. fulvipes* by earlier British authors. Fore wing length: ♂♀ 10-11 mm (Fig. 325). Common in England, Wales and mainland Scotland; present in Ireland; streams and rivers. Most of Europe, less common in north. Flight period: (April) May-August. ♂ genitalia with long processes on segment X, dorsal carina parallel-sided, aedeagus with moderate-sized teeth (Fig. 326); ♀ genitalia with very small dorsal lobe that does not meet ventral lobe, clasper receptacle with small round opening (Fig. 327).

325

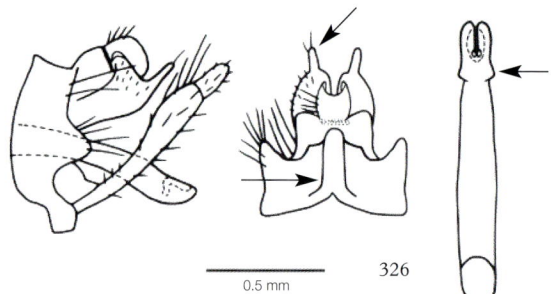

0.5 mm

326

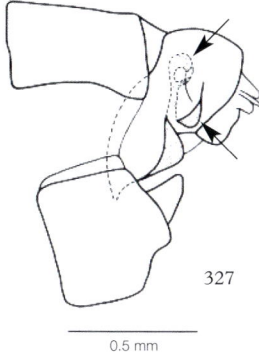

Figures 325-327. *Hydropsyche instabilis.* 325 wing pattern;
326 male genitalia lateral and dorsal, aedeagus ventral; 327 female genitalia lateral

## *Hydropsyche siltalai* Döhler, 1963

Sometimes misidentified as *H. instabilis* by previous authors. Fore wing length: ♂♀ 10-12 mm (Figs 328, 329). Common throughout most of England, Wales and Scotland; present in Ireland; streams and rivers. Throughout Europe. Flight period: June-September. ♂ genitalia similar to *H. instabilis* but dorsal carina on segment X either triangular or widest in the centre, never parallel-sided, aedeagus with very small teeth (Fig. 330); ♀ genitalia with very long ventral lobe which touches or overlaps dorsal lobe, clasper receptacle with small keyhole-shaped opening (Fig. 331).

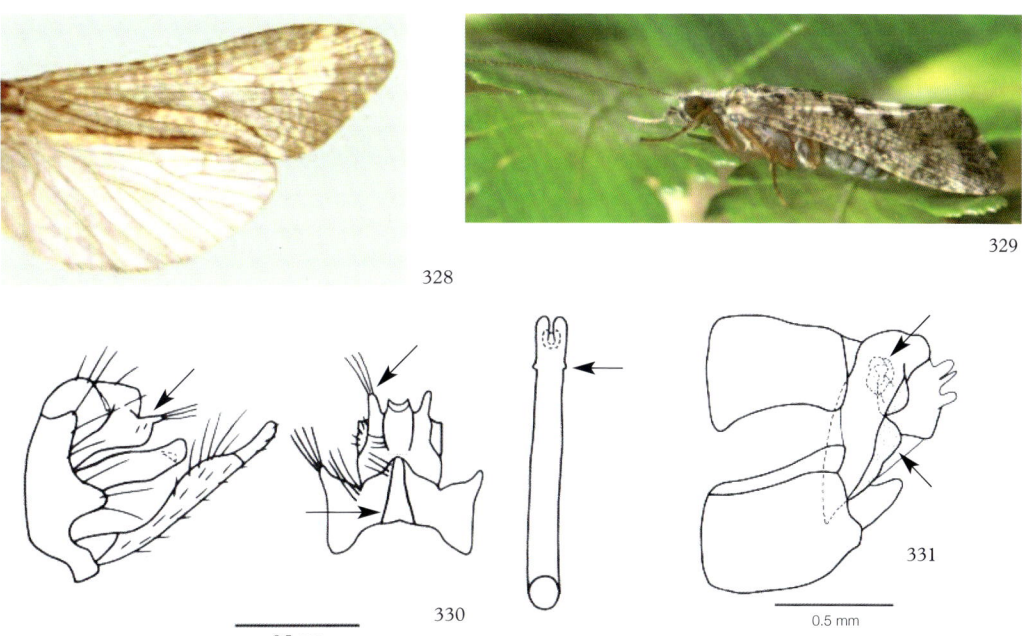

Figures 328-331. *Hydropsyche siltalai.* 328 wing pattern; 329 live specimen;
330 male genitalia lateral and dorsal, aedeagus ventral; 331 female genitalia lateral

85

## *Hydropsyche saxonica* McLachlan, 1884

First recorded in Britain by Grensted (1943). Fore wing length: ♂♀ 11-13 mm (Fig. 332). Very local in England, Wales and SW. Scotland; present in Ireland; streams. Mainly central and northern Europe. Flight period: June-August. ♂ genitalia with short processes on segment X, dorsal carina broadly triangular, teeth on aedeagus moderate-sized and relatively far back from apex (Fig. 333); ♀ genitalia with small dorsal and ventral lobes that almost touch, clasper receptacle small with irregular opening, often with pointed upper corners (Fig. 334).

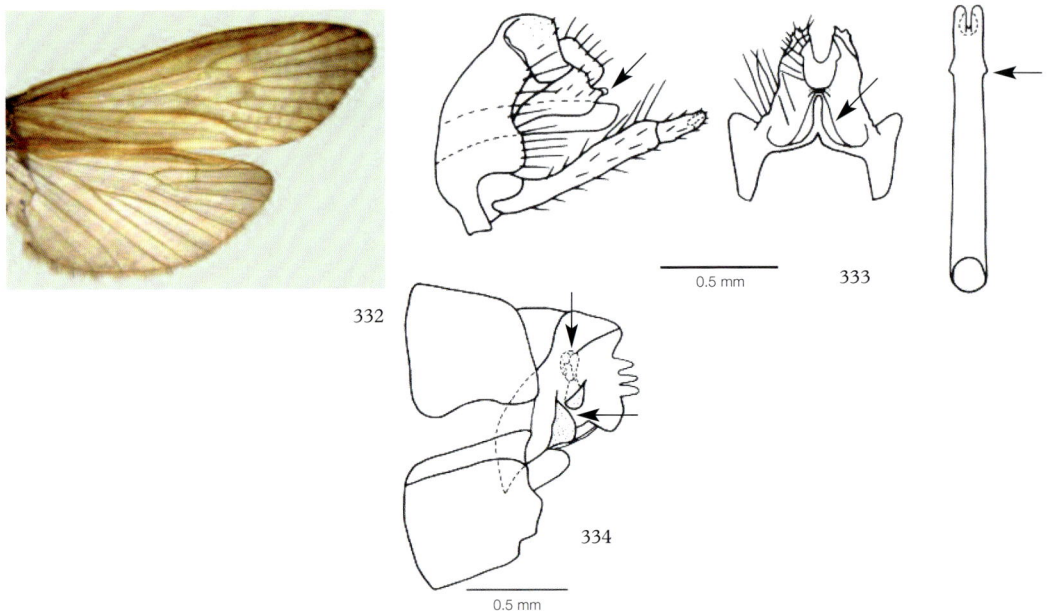

Figures 332-334. *Hydropsyche saxonica*. 332 wing pattern; 333 male genitalia lateral and dorsal, aedeagus ventral; 334 female genitalia lateral

## *Hydropsyche fulvipes* (Curtis, 1834)

Fore wing length: ♂♀ 9-11 mm (Fig. 335): a large and dark brown species, though the wings are not unicolorous. Very local in England, Wales and mainland Scotland; present in Ireland; spring-fed streams. Throughout much of Europe except extreme north and south. Flight period: May-September. ♂ genitalia with long processes on segment X, very narrow dorsal carina, teeth on aedeagus very small (Fig. 336); ♀ genitalia with very small dorsal lobe, small clasper receptacle with clear canal leading to it (Fig. 337).

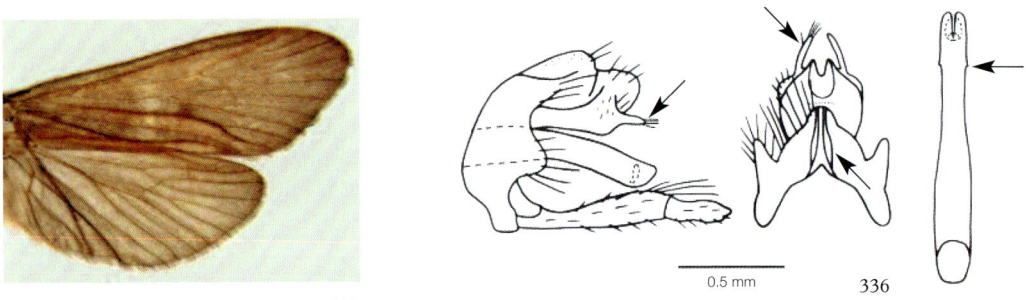

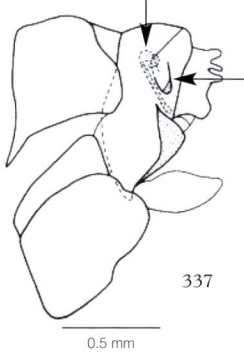

337

0.5 mm

Figures 335-337. *Hydropsyche fulvipes.* 335 wing pattern;
336 male genitalia lateral and dorsal, aedeagus ventral; 337 female genitalia lateral

## *Hydropsyche pellucidula* (Curtis, 1834)

Fore wing length: ♂♀ 9-14 mm (Fig. 338). Very common in England, Wales and Scotland; present in Ireland; streams and rivers. Throughout Europe. Flight period: May-September. ♂ genitalia with no processes on segment X, dorsal carina narrow, teeth on aedeagus very large and clearly triangular (Fig. 339); ♀ genitalia with elongate club-shaped opening to clasper receptacle (Fig. 340).

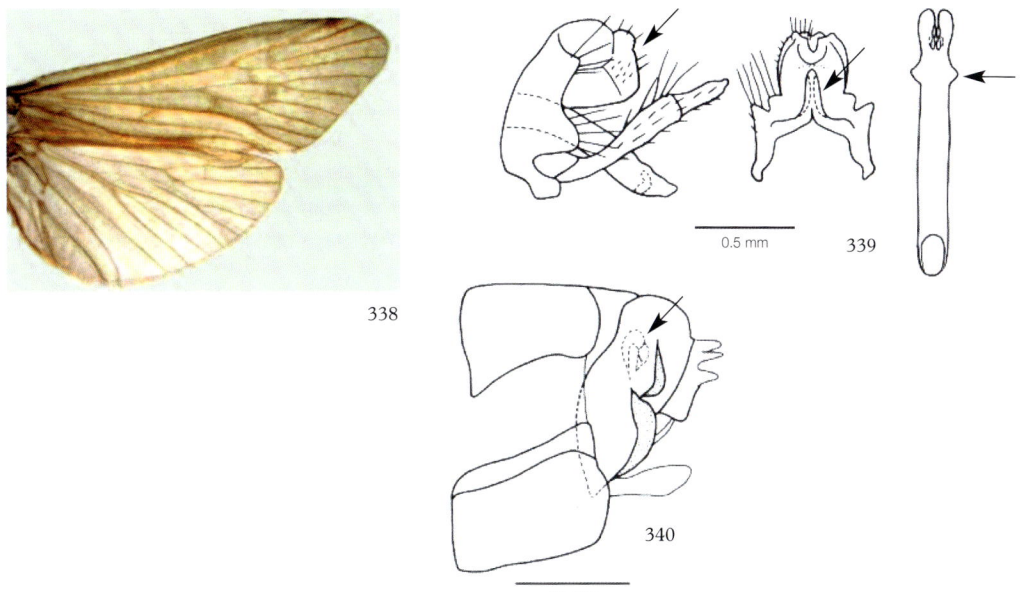

338

0.5 mm  339

340

0.5 mm

Figures 338-340. *Hydropsyche pellucidula.* 338 wing pattern;
339 male genitalia lateral and dorsal, aedeagus ventral; 340 female genitalia lateral

## *Hydropsyche contubernalis* McLachlan, 1865
= *ornatula* McLachlan, 1878

Listed as *H. ornatula* in Mosely (1939). Fore wing length: ♂ 7-11 mm, ♀ 9-14 mm (Fig. 341): a large white spot on hind margin of fore wing. England, Wales and mainland Scotland; present in Ireland; rivers. Most of Europe. Flight period: (April) May-September. ♂ genitalia with no processes on segment X, dorsal carina broad and short, aedeagus with no teeth (Fig. 342); ♀ genitalia with a clear gap between the dorsal and ventral lobes, clasper rectacle quite large and asymmetrical (Fig. 343).

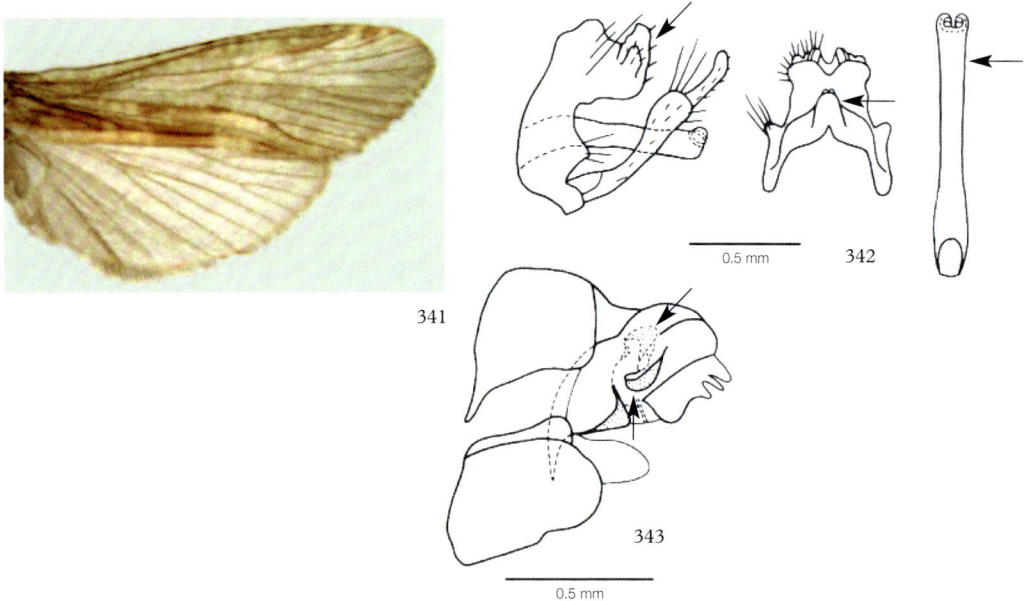

Figures 341-343. *Hydropsyche contubernalis*. 341 wing pattern;
342 male genitalia lateral and dorsal, aedeagus ventral; 343 female genitalia lateral

## *Hydropsyche bulgaromanorum* Malicky, 1977
= *guttata*; misidentified by some authors

Previously misidentified as *H. guttata* (a central European species not occurring in Britain) in Mosely (1939) and Macan (1973) and other works. See Malicky (1984b). This is a BAP species: the Scarce Grey Flag. Fore wing length: ♂♀ 9-10 mm (Fig. 344). S. England, previously thought to be extinct in Britain but recently rediscovered in W. Sussex (Drake & Willo, 2009); no records from Ireland; lower reaches of large rivers. Most of Europe (many early records may be confused with *guttata*). Flight period: August-September. ♂ genitalia with no processes on segment X, dorsal carina narrow, aedeagus with small bulge behind apex but no teeth (Fig. 345); ♀ genitalia with ventral lobe triangular and acutely pointed at apex, clasper receptacle large and bell-shaped (Fig. 346).

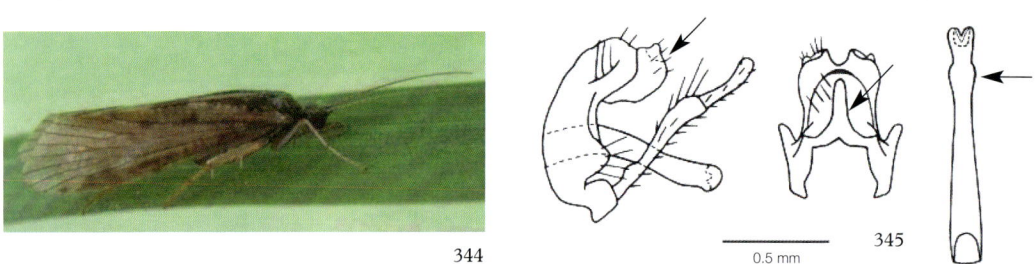

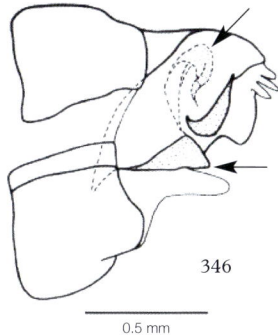

Figures 344-346. *Hydropsyche bulgaromanorum*. 344 live specimen [Photo: Peter Barnard]; 345 male genitalia lateral and dorsal, aedeagus ventral; 346 female genitalia lateral

## *Hydropsyche exocellata* **Dufour, 1841**

Fore wing length: ♂ 8-9 mm, ♀ 9-10 mm (Fig. 347). The male eyes (Fig. 348) are noticeably larger than those of the female. Old records from S. England, but now considered extinct; no records from Ireland; lower reaches of large rivers, especially R. Thames. Western and central Europe. Flight period: May-September. ♂ genitalia with no processes on segment X, dorsal carina short and triangular, aedeagus with bulge behing apex but no teeth (Fig. 349); ♀ genitalia with small triangular ventral lobe, clasper receptacle moderately large with simple rounded opening (Fig. 350).

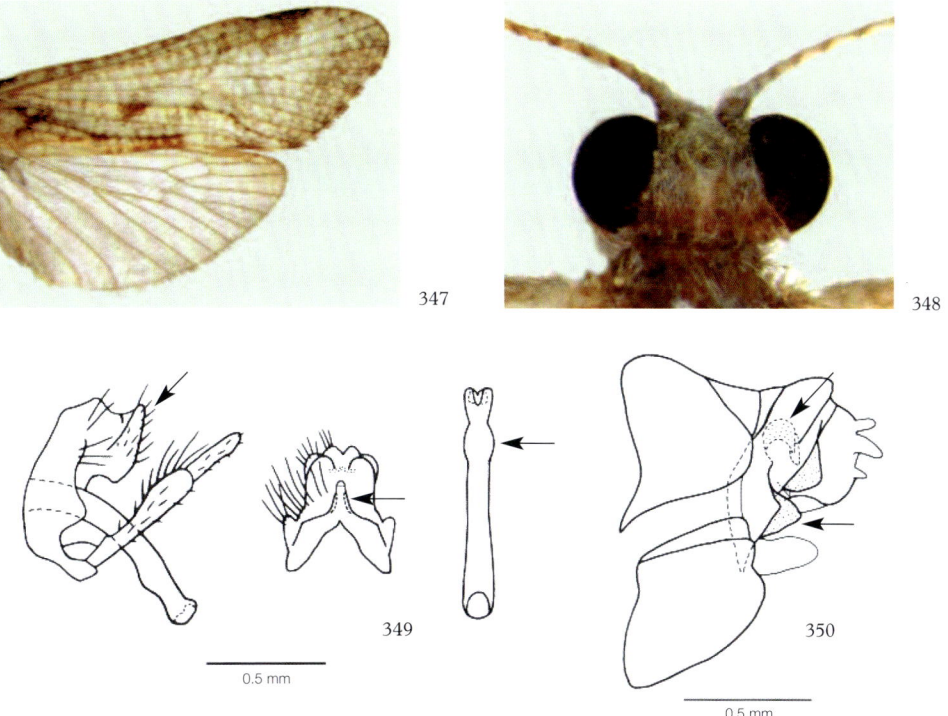

Figures 347-350. *Hydropsyche exocellata*. 347 wing pattern; 348 male eyes; 349 male genitalia lateral and dorsal, aedeagus ventral; 350 female genitalia lateral

### *Hydropsyche angustipennis* (Curtis, 1834)

Fore wing length: ♂ 8-10 mm, ♀ 9-11 mm (Fig. 351); a small, dark brown unicolorous species. Common in lowland but very local elsewhere in England, Wales and mainland Scotland; present in Ireland; streams and rivers, especially the exits of ponds and lakes. Throughout Europe. Flight period: (April) May-September. ♂ genitalia with no processes on segment X, dorsal carina broad and parallel-sided, aedeagus with large bulge behind apex but no teeth (Fig. 352); ♀ genitalia with very small and round clasper receptacle, dorsal and ventral lobes of similar small size (Fig. 353).

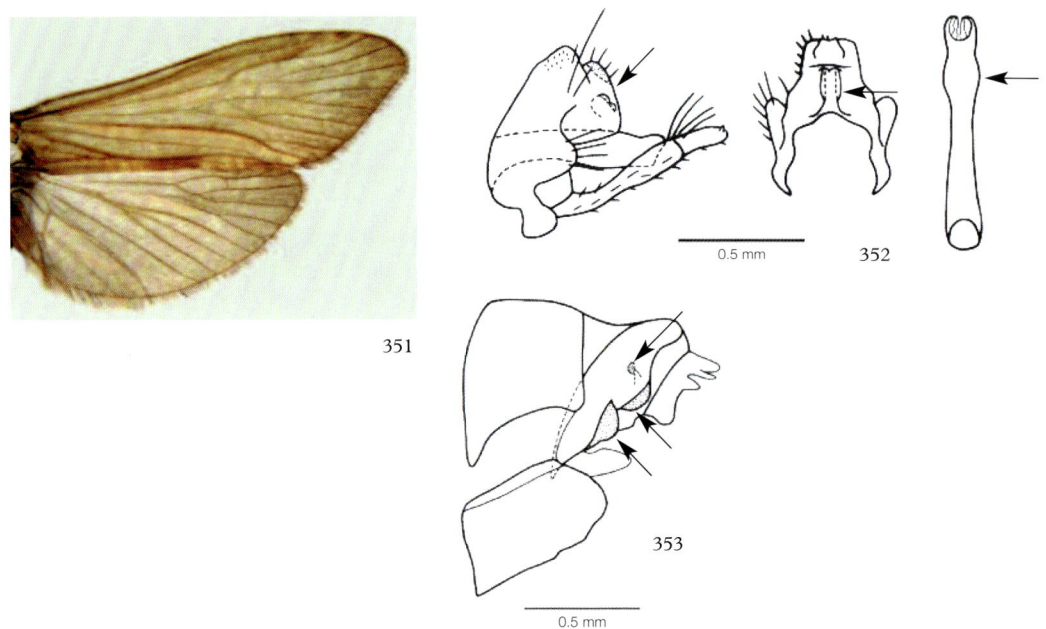

Figures 351-353. *Hydropsyche angustipennis.* 351 wing pattern;
352 male genitalia lateral and dorsal, aedeagus ventral; 353 female genitalia lateral

# Family PHRYGANEIDAE (5 genera, 10 species)

Fore wing described under each genus below; discoidal closed in both wings. Spur formula 2.4.4 (fewer in *Agrypnetes crassicornis*). Ocelli present. Male maxillary palps with only four segments.

The current nomenclature is based largely on Wiggins' (1998) revision of the family.

The following key to genera leads partly to generic groups, because some genera can be separated only by genitalic characters, and these are easily distinguished from the following drawings of each species. Wing venation cannot readily be used to separate genera because of widespread sexual dimorphism.

## Key to genera of Phryganeidae

1. Mid tarsi flattened and fringed; spurs fewer than 2.4.4; fore wing uniformly pale brown with even paler veins ...................................................................... *Agrypnetes* (p. 92)

- Mid tarsi not flattened or fringed; spur formula 2.4.4; fore wing patterned or mid to dark brown ................................................................................................................ 2

2. Males (maxillary palps 4-segmented) ............................................................. 3

- Females (maxillary palps 5-segmented) ........................................................... 5

3. Ventral margin of sternite IX with right-angled posterior corner in lateral view (Fig. 354) .................................................................................................... *Hagenella* (p. 98)

- Ventral margin of sternite IX simple (Fig. 355) ............................................ 4

354

355

0.5 mm

4. Fore wing length greater than 17 mm ................................................. *Phryganea* (p. 100)

- Fore wing length less than 17 mm .................................................................................. .............................................. *Agrypnia* (p. 94), *Trichostegia* (p. 99) and *Oligotricha* (p. 100)

5. Subgenital plate narrowed apically (Fig. 356) ................................................. 6

- Subgenital plate not narrowed but truncate (Fig. 357) ...................................................... .................................................................. *Hagenella* (p. 98) and *Oligotricha* (p. 100)

356

357

6. Subgenital plate with no processes or lobes (Fig. 358) ........................ *Trichostegia* (p. 99)

- Subgenital plate with conspicuous processes (Figs 359-361) ..........................................
...................................................................... *Agrypnia* (p. 94) and *Phryganea* (p. 100)

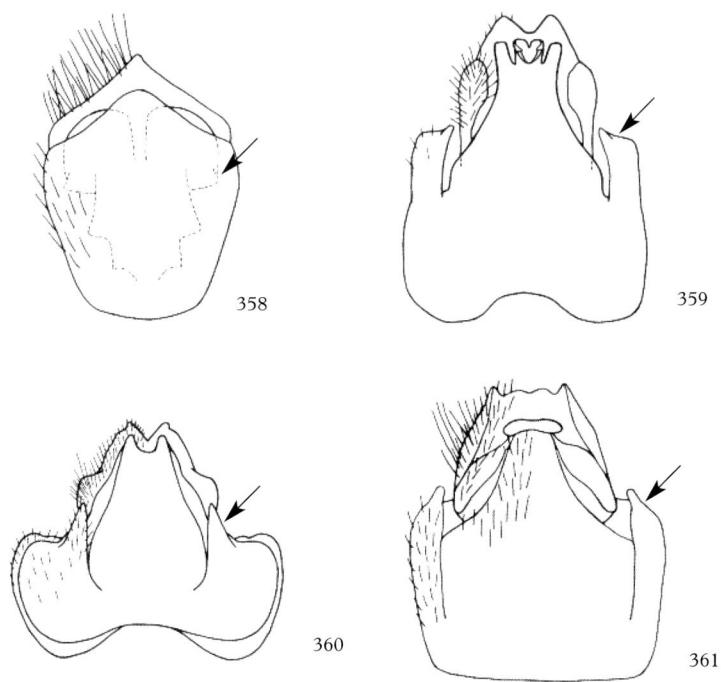

# Genus AGRYPNETES McLachlan, 1876

Fore wing with forks 1, 2, 3 and 5 present, hind wing with forks 1, 2 and 5 (Fig. 362). Spur formula 1.2.2 or possibly 1.3.3 (difficult to distinguish in some specimens).

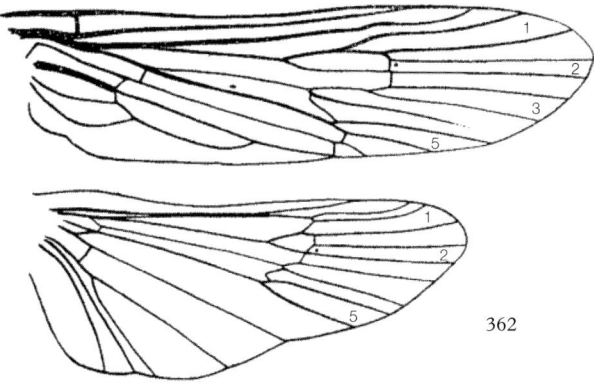

Figure 362. *Agrypnetes crassicornis.* Wing venation

## *Agrypnetes crassicornis* (McLachlan, 1876)

Some authors have regarded *Agrypnetes* as synonym of *Agrypnia*, but the marked differences in larval, pupal and adult characters (Wiggins, 1998) seem to justify its retention as a distinct genus. However, the discovery of a second species of *Agrypnetes* in Mongolia (Morse & Chuluunbat, 2007) suggests that the two genera may be more closely related than Wiggins believed. Not recorded as British until Kimmins (1952). Fore wing length: ♂ 10-12 mm, ♀ 15-17 mm (Figs 362-364); pale brown with conspicuously paler veins but no darker markings. Known only from Malham Tarn, Yorkshire (see Ross, 2008); no records from Ireland; alkaline or saline lakes. Northern Europe only, especially in brackish water (Gullefors, 2005). Flight period: June-July; though it is really a flightless species that skims across the surface of the water. ♂ genitalia as in Fig. 365; ♀ genitalia as in Fig. 366.

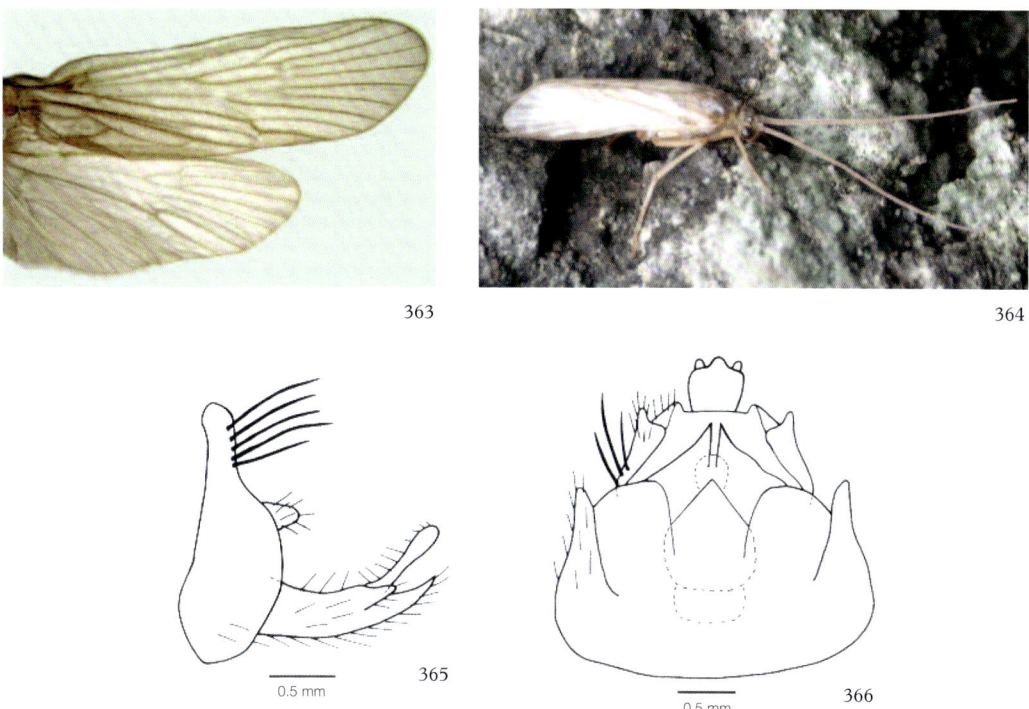

363

364

365

0.5 mm

366

0.5 mm

Figures 363-366. *Agrypnetes crassicornis.* 363 wing pattern; 364 live specimen; 365 male genitalia lateral, aedeagus omitted; 366 female genitalia ventral

# Genus AGRYPNIA Curtis, 1835

Fore wing with forks 1, 2, 3 and 5 present in male, but 1 to 5 in female; hind wing with forks 1, 2 and 5 in male, 1, 2, 3 and 5 in female (fork 3 sometimes absent) (Figs 375, 376). About six species in Europe, four of which occur in Britain.

## *Agrypnia varia* (Fabricius, 1793)

Listed as *Phryganea varia* in Mosely (1939) and Macan (1973); also known as *Dasystegia varia*. This is the angler's Speckled Peter. Fore wing length: ♂ 12-17 mm ♀ 15-18 mm (Figs 367, 368) fore wing strongly marked. Common throughout Britain, but less frequent than *A. obsoleta* in upland acid waters; present in Ireland; ponds, lakes. Throughout Europe. Flight period: June-September. ♂ genitalia with dorsal lobe of clasper bearing fringe of dark hairs (Fig. 369); ♀ genitalia subgenital plate long and approximately triangular (Fig. 370).

Figures 367-370. *Agrypnia varia*. 367 wing pattern; 368 live specimen; 369 male genitalia lateral, aedeagus omitted; 370 female genitalia ventral

94

## *Agrypnia obsoleta* (Hagen, 1864)

Listed as *Phryganea obsoleta* in Mosely (1939) and Macan (1973); also known as *Dasystegia obsoleta*. This is the angler's Dark Peter. Fore wing length: ♂ 9-15 mm, ♀ 13-16 mm (Figs 371, 372) wing markings less contrasting than in *A. varia*. Wales, N. England and Scotland, with a few records from SW. England; present in Ireland; usually upland pools and bogs. Mainly central and northern Europe. Flight period: June-September. ♂ genitalia with apex of clasper divided into two finger-like lobes (Fig. 373); ♀ genitalia similar to *A. varia* but subgenital plate shorter and broader (Fig. 374).

372

371

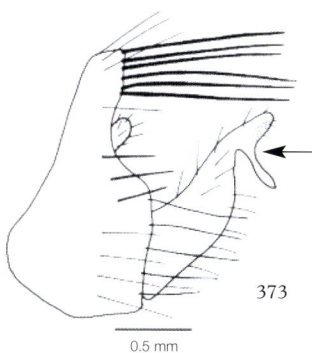

373

0.5 mm

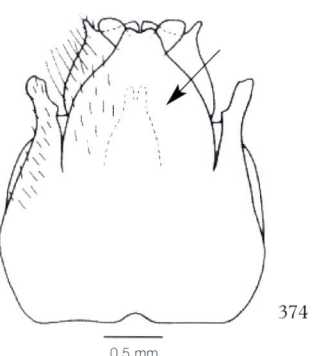

374

0.5 mm

Figures 371-374. *Agrypnia obsoleta*. 371 wing pattern; 372 live specimen; 373 male genitalia lateral, aedeagus omitted; 374 female genitalia ventral

## *Agrypnia picta* Kolenati, 1848

Fore wing length: ♂♀ 13-15 mm (Figs 375-377). This species is easily confused with species of Limnephilidae if not examined carefully, because of the wing shape and rather uniformly coloured fore wing. Fork 3 usually absent in female hind wing. Known only from two old records (the unlikely extremes of London and Shetland), which cannot be verified and may well be incorrect; no records from Ireland; lake fens. Mainly northern Europe. Flight period: June-August. ♂ genitalia with club-shaped apical segment to clasper bearing dark hairs (Fig. 378); ♀ genitalia as in Fig. 379.

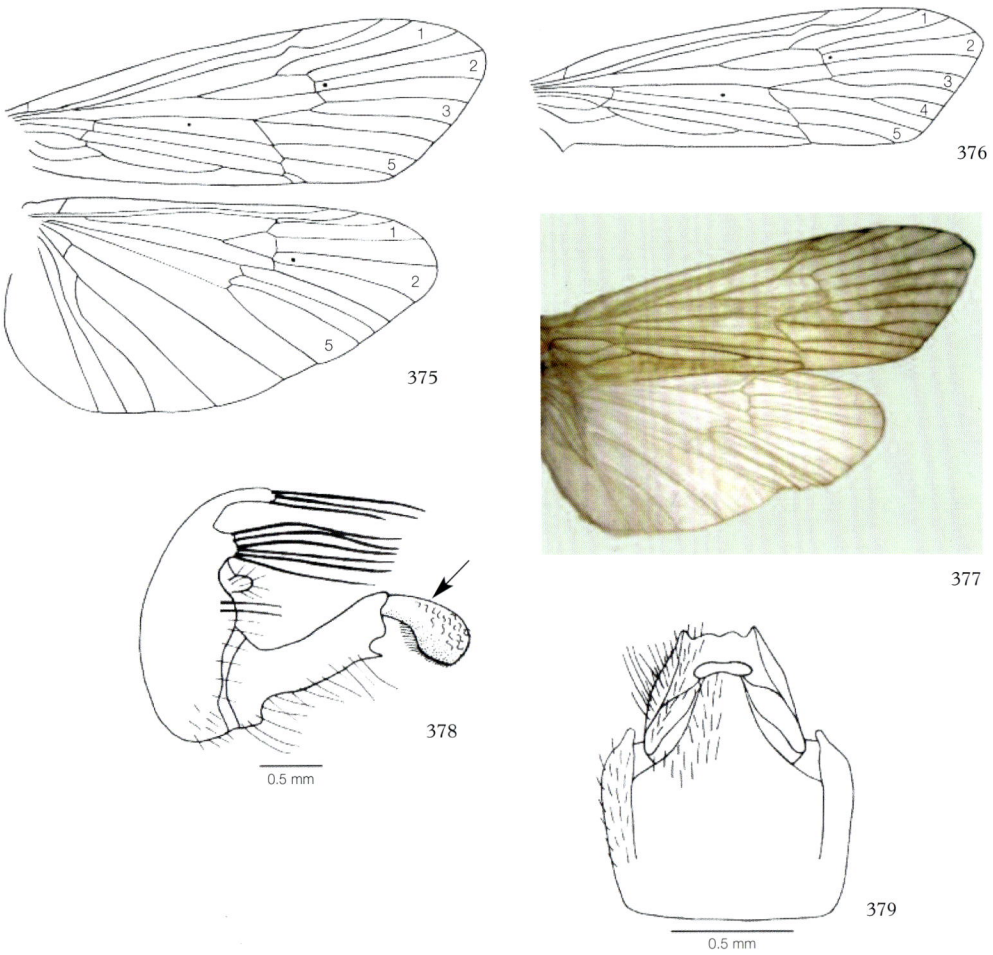

Figures 375-379. *Agrypnia picta*. 375 male wing venation; 376 female fore wing; 377 wing pattern; 378 male genitalia lateral, aedeagus omitted; 379 female genitalia ventral

## *Agrypnia pagetana* **Curtis, 1835**

Fore wing length: ♂♀ 10-15 mm (Figs 380-382). Like *A. picta*, this species superficially resembles a limnephilid and has a very uniform colour. Fork 3 usually absent in female hind wing. Local in southern central and eastern England, and southern and eastern Scotland; present in Ireland; ponds, lakes and canals. Throughout much of Europe except extreme south. Flight period: May-September. ♂ genitalia with apex of clasper divided into two elongate lobes (Fig. 383); ♀ genitalia as in Fig. 384.

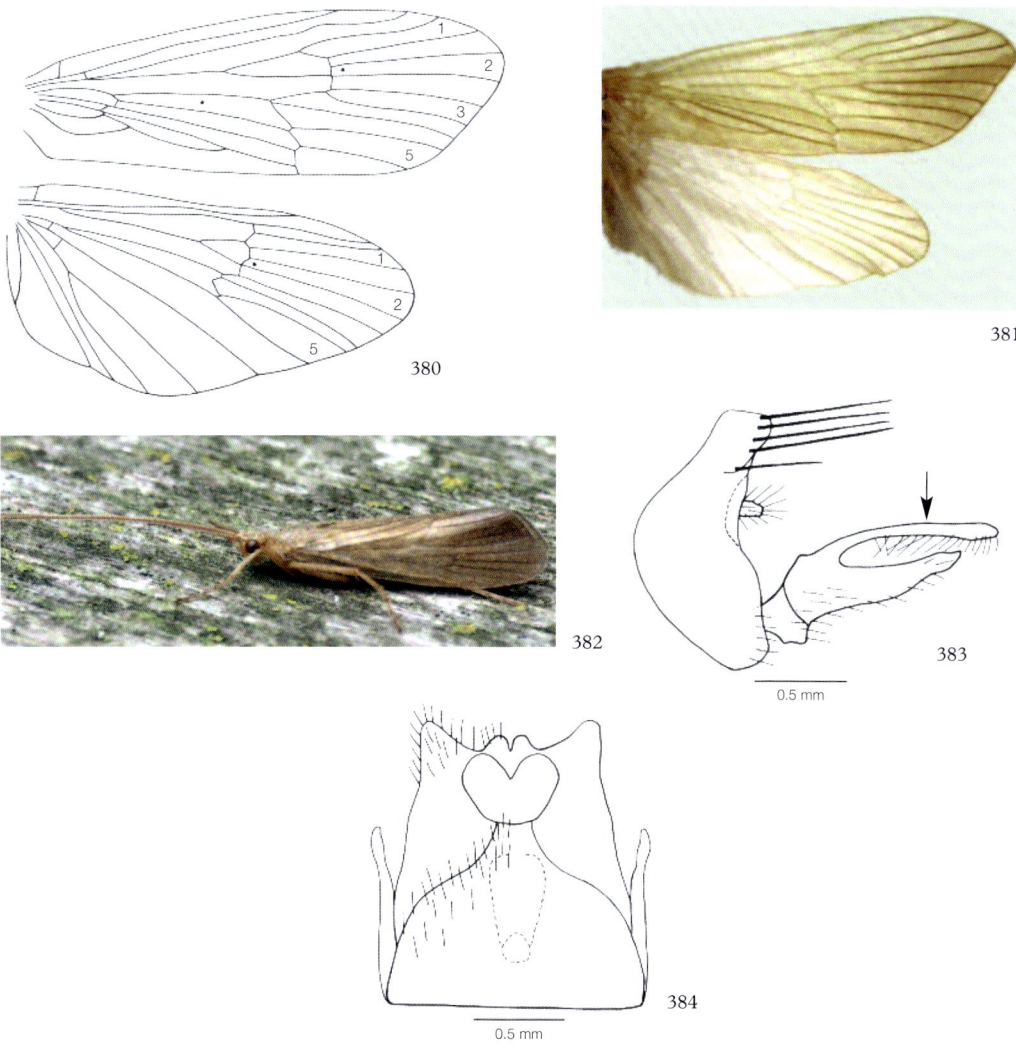

Figures 380-384. *Agrypnia pagetana*. 380 male wing venation; 381 wing pattern; 382 live specimen; 383 male genitalia lateral, aedeagus omitted; 384 female genitalia ventral

# Genus HAGENELLA Martynov, 1924

Fore wing with forks 1, 2, 3 and 5 present in both sexes; hind wing with forks 1, 2 and 5 in male, 1, 2, 3 and 5 in female. *H. clathrata* is the only European species in this genus.

## *Hagenella clathrata* (Kolenati, 1848)

Listed as *Neuronia clathrata* in Mosely (1939) and *Oligotrichia* [sic] *clathrata* in Macan (1973). This is a BAP species: the Window Winged Sedge. Fore wing length: ♂ 11-15 mm, ♀ 12-15 mm (Fig. 385); very distinctive wing markings. Rare, with recent records only from Surrey, NW Midlands and adjoining parts of Wales, and Aviemore region of Scotland; no records from Ireland; tiny pools on raised bogs and heaths. Mainly central and northern Europe. Flight period: June-July. ♂ genitalia as in Fig. 386; ♀ genitalia as in Fig. 387.

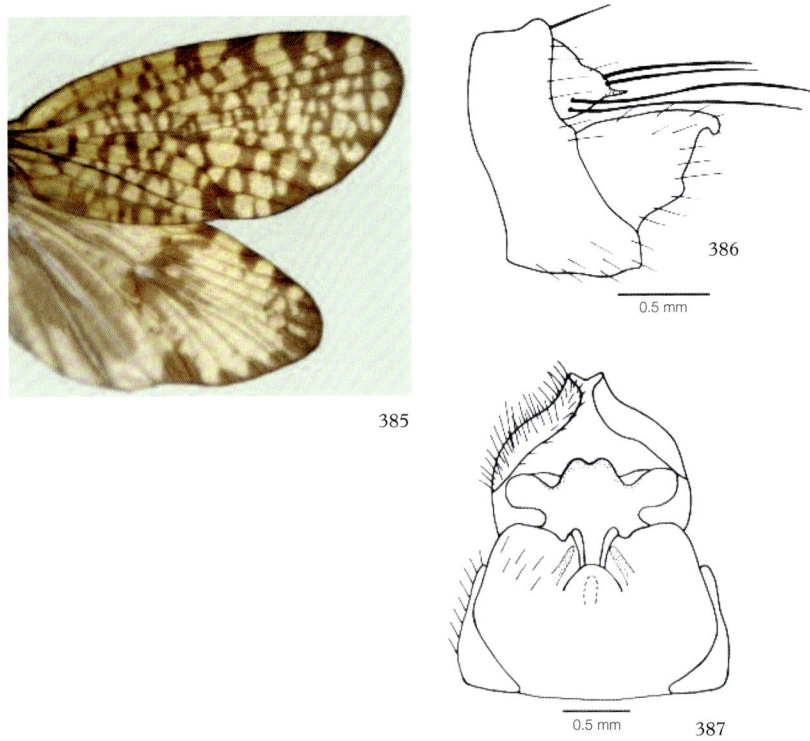

Figures 385-387. *Hagenella clathrata*. 385 wing pattern;
386 male genitalia lateral, aedeagus omitted; 387 female genitalia ventral

# Genus TRICHOSTEGIA Kolenati, 1848

Fore wing with forks 1, 2, 3 and 5 in both sexes; hind wing with forks 1, 2 and 5 (Fig. 388). *T. minor* is the only European species.

## *Trichostegia minor* (Curtis, 1834)

Listed as *Nannophryganea minor* in Mosely (1939). Fore wing length: ♂♀ 9-12 mm (Figs 388, 389); distinctive wing pattern resembling a small specimen of *Phryganea*. Common in England, less common in Wales, one record for SW. Scotland; no records from Ireland; small temporary pools and ditches in woods and fens. Throughout most of Europe except extreme south. Flight period: June-July. ♂ genitalia as in Fig. 390; ♀ genitalia as in Fig. 391.

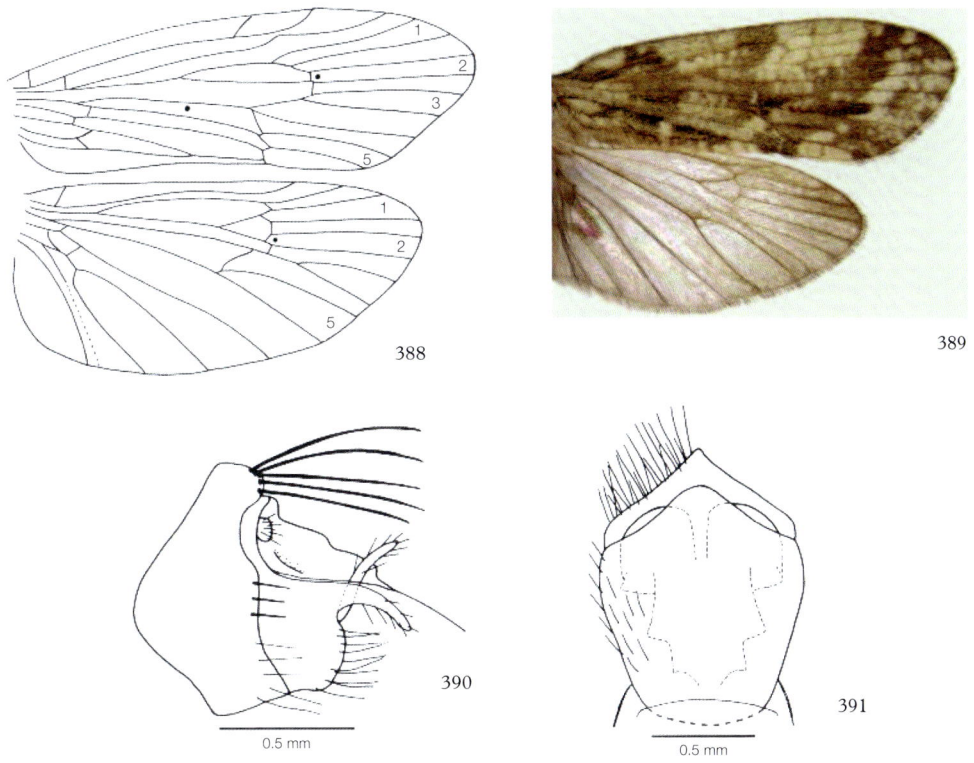

Figures 388-391. *Trichostegia minor.* 388 wing venation; 389 wing pattern; 390 male genitalia lateral, aedeagus omitted; 391 female genitalia ventral

## Genus OLIGOTRICHA Rambur, 1842

Fore wing with forks 1, 2, 3 and 5 present in both sexes; hind wing with forks 1, 2 and 5 in male, 1, 2, 3 and 5 in female (Figs 392, 393). Two species in Europe, with one in Britain.

### *Oligotricha striata* (Linnaeus, 1758)
= *ruficrus* (Scopoli, 1763)

Listed as *Neuronia ruficrus* in Mosely (1939) and *Oligotrichia* [sic] *ruficrus* in Macan (1973). Fore wing length: ♂♀ 13-16 mm (Figs 392-394); living specimens are uniformly black (Fig. 394) though this colour fades to brown after death; antennae clearly serrate. Local in S. and N. England, Wales, NW Midlands and mainland Scotland; present in Ireland; pools and ditches with abundant *Sphagnum* moss. Most of Europe except extreme south. Flight period: May-July. ♂ genitalia as in Fig. 395; ♀ genitalia as in Fig. 396.

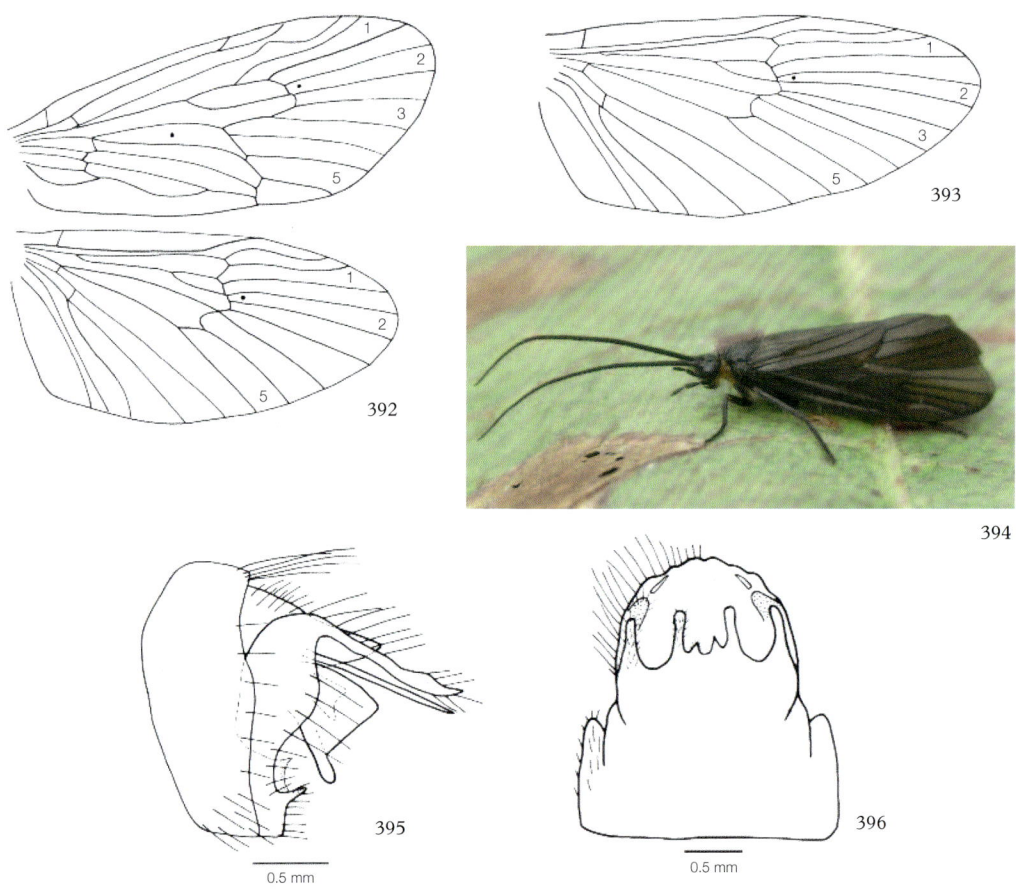

Figures 392-396. *Oligotricha striata*. 392 male wing venation; 393 female hind wing; 394 live specimen; 395 male genitalia lateral, aedeagus omitted; 396 female genitalia ventral

## Genus PHRYGANEA Linnaeus, 1758

Fore wing with forks 1, 2, 3 and 5 present in male, 1 to 5 in female; hind wing with forks 1, 2 and 5 in male, 1, 2, 3 and 5 in female (Figs 397, 398). These are the angler's Great (or Large) Red Sedges, or Murraghs. Both European species are found in Britain.

## *Phryganea grandis* Linnaeus, 1758

Fore wing length: ♂ 18-25 mm, ♀ 20-28 mm (Figs 397-401); female usually with an almost unbroken dark bar along centre of fore wing. Throughout Britain, but with few records from Wales and Scotland; present in Ireland; ponds, lakes, canals and slow rivers. Throughout most of Europe except extreme south. Flight period: June-August. Adults are commonly caught in light-traps though there are few larval records from many areas. ♂ genitalia with strongly incurved claspers (Fig. 402); ♀ genitalia with narrow pointed lateral processes on subgenital plate (Fig. 403).

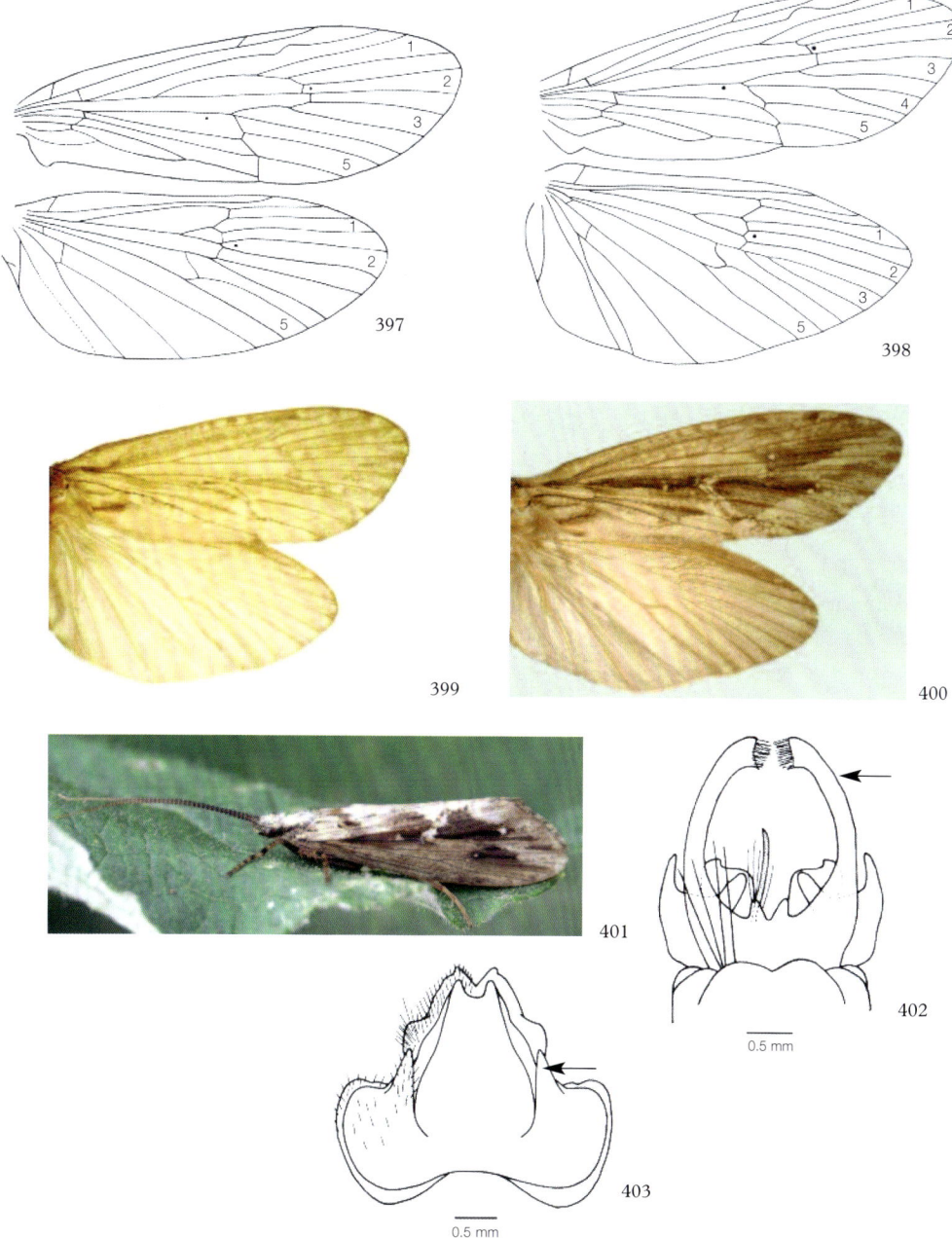

Figures 397-403. *Phryganea grandis.* 397 male wing venation; 398 female wing venation; 399 male wing pattern; 400 female wing pattern; 401 female live specimen; 402 male genitalia dorsal, aedeagus omitted; 403 female genitalia ventral

## *Phryganea bipunctata* **Retzius, 1783**
= *striata*; misidentified by some authors

Listed as *P. striata* in Mosely (1939) and Macan (1973). Fore wing length: ♂ 18-25 mm, ♀ 18-26 mm (Figs 404-406): in the female the two sections of the dark central wing bar are usually well separated by a pale area, though this is not a reliable separation from *P. grandis*. Common throughout Britain, less common in Scotland; present in Ireland; ponds, lakes, canals and slow rivers. Throughout most of Europe except extreme south. Flight period: May-July. ♂ genitalia with converging claspers that are almost straight (Fig. 407); ♀ genitalia with broad and blunt lateral processes on subgenital plate (Fig. 408).

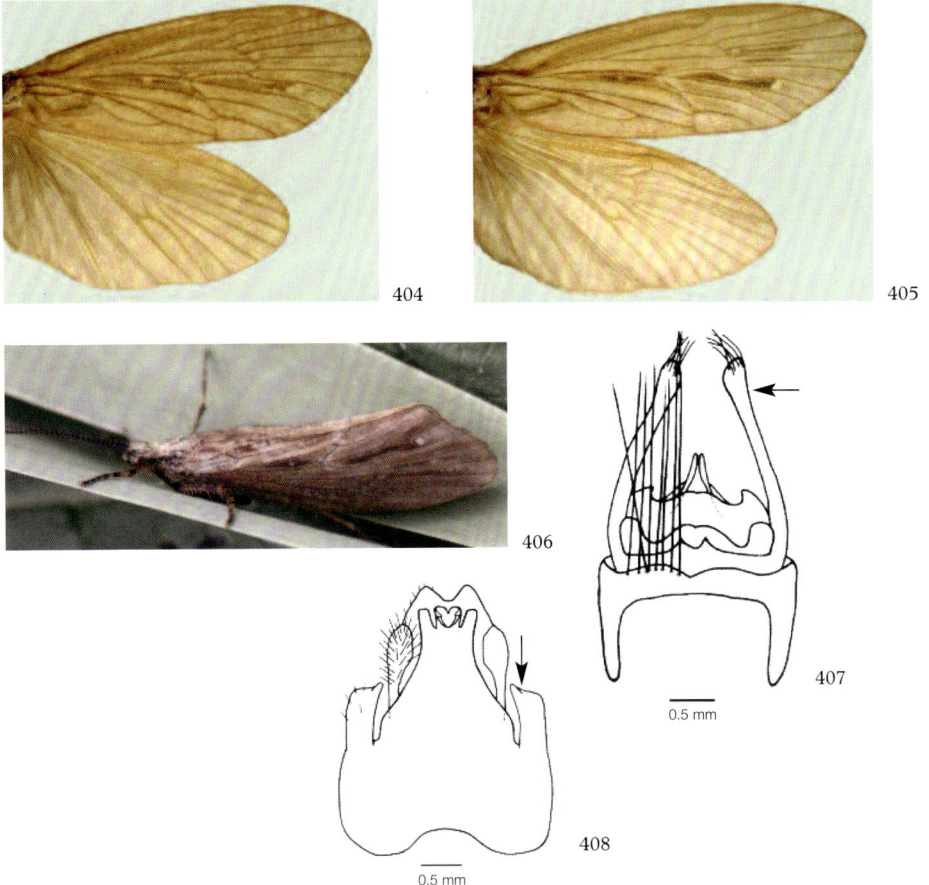

Figures 404-408. *Phryganea bipunctata*. 404 male wing pattern; 405 female wing pattern; 406 female live specimen; 407 male genitalia dorsal, aedeagus omitted; 408 female genitalia ventral

# Family BRACHYCENTRIDAE (1 genus, 1 species)

Fore wing with forks 1, 2, 3 and 5 present in male, 1 to 5 in female; hind wing with forks 1 and 5 in male, 1, 2, 3 and 5 in female. Discoidal cell closed in fore wing, open in hind wing (Figs 409, 410). Spur formula 2.3.3. Ocelli absent. Male maxillary palps with only three segments. *Brachycentrus subnubilis* is the only British representative of this family; around three more occur in Europe.

# Genus BRACHYCENTRUS Curtis, 1834

## *Brachycentrus subnubilus* Curtis, 1834

This is the angler's Grannom or Greentail. Fore wing length: ♂ 7-9 mm, ♀ 10-14 mm (Figs 409-412); a distinctive pattern of yellowish spots on a grey background quickly fades after death. Throughout England, Wales and mainland Scotland, but possibly declining in some rivers in southern England; no records from Ireland; fast rivers and streams. Throughout most of Europe. Flight period: March-June (July). A day-flying species, not attracted to light. ♂ genitalia as in Fig. 413; ♀ genitalia with large concave end in which the protruded egg-mass is carried (Fig. 414).

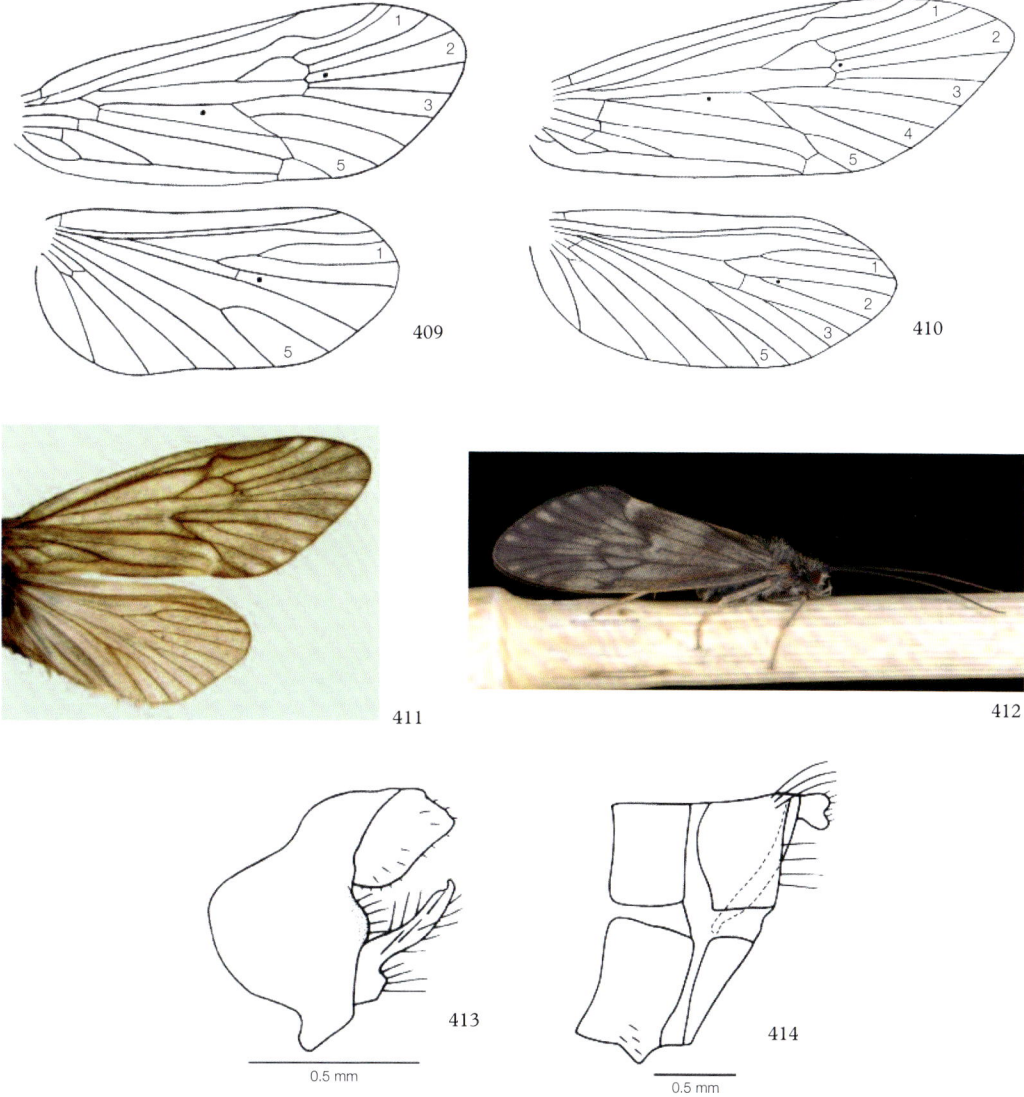

Figures 409-414. *Brachycentrus subnubilus*. 409 male wing venation; 410 female wing venation; 411 wing pattern; 412 live specimen [Photo: Stuart Crofts]; 413 male genitalia lateral; 414 female genitalia lateral

# Family GOERIDAE (2 genera, 3 species)

Fore wing with forks 1, 2, 3 and 5 in both wings; discoidal cell closed in fore wing, open in hind wing (Fig. 417). Spur formula 2.4.4. Ocelli absent. Male maxillary palps with three segments that are highly modified.

## Key to genera of Goeridae

1. Fore wing brownish in life; the area between Cu and A1 expanded just behind apical fork 5 (Fig. 415) .................................................................................... *Goera* (p. 104)

- Fore wing black in life; the area between Cu and A1 not expanded (Fig. 416) ................. .................................................................................................. *Silo* (p. 105)

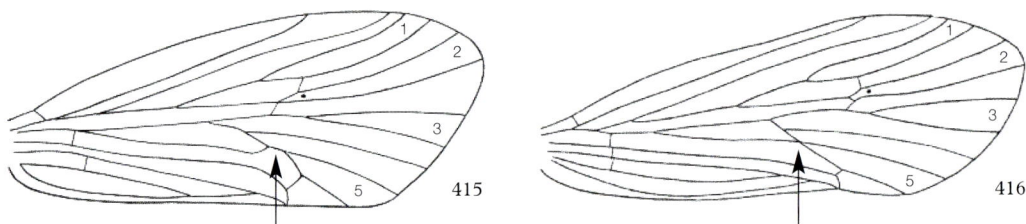

## Genus GOERA Stephens, 1829

*G. pilosa* is the only European species.

### *Goera pilosa* (Fabricius, 1775)

This is the angler's Medium Sedge. Fore wing length: ♂ 8-10 mm, ♀ 9-12 mm (Figs 417-419); fore wing brownish yellow with yellow hairs in life. Throughout Britain, though most abundant in the chalk areas of SE. England; present in Ireland; stony streams, rivers and lakes. Throughout Europe. Flight period: May-September. ♂ genitalia as in Fig. 420; ♀ genitalia as in Fig. 421.

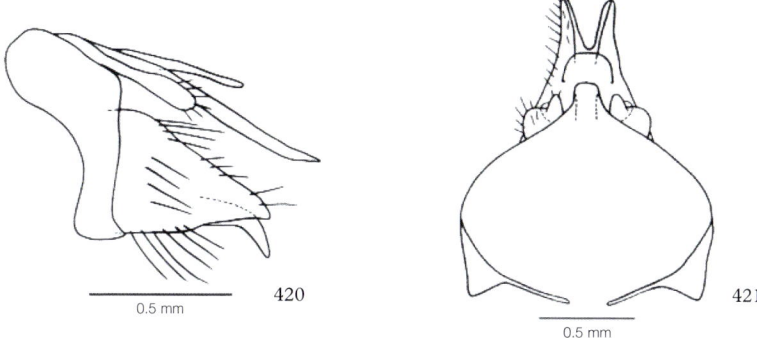

Figures 417-421. *Goera pilosa.* 417 wing venation; 418 wing pattern; 419 live specimen;
420 male genitalia lateral, aedeagus omitted; 421 female genitalia ventral

# Genus SILO Curtis, 1833

These are the angler's Black Sedges. Two British species out of around 10 in Europe.

## *Silo nigricornis* (Pictet, 1834)

Fore wing length: ♂ 6-11 mm, ♀ 7-12 mm (Figs 422-425); yellow hairs on a dark background, male usually darker than female, though colours fade after death. Male with angular groove of scent hairs on hind wing. Base of fork 1 in fore wing overlaps more than half of discoidal cell (Fig. 422). Common in the chalk areas of SE. England, local in SW. and N. England, Wales and Scotland as far north as Aviemore; present in Ireland; stony rivers and streams. Mainly southern and central Europe. Flight period: May-September. ♂ genitalia with short superior appendages (Fig. 426); ♀ genitalia with short and broad segment VIII bearing rounded lobes (Fig. 427). Note that the female can usually be separated from that of *S. pallipes* by the very long fork 1 in the hind wing (Fig. 423).

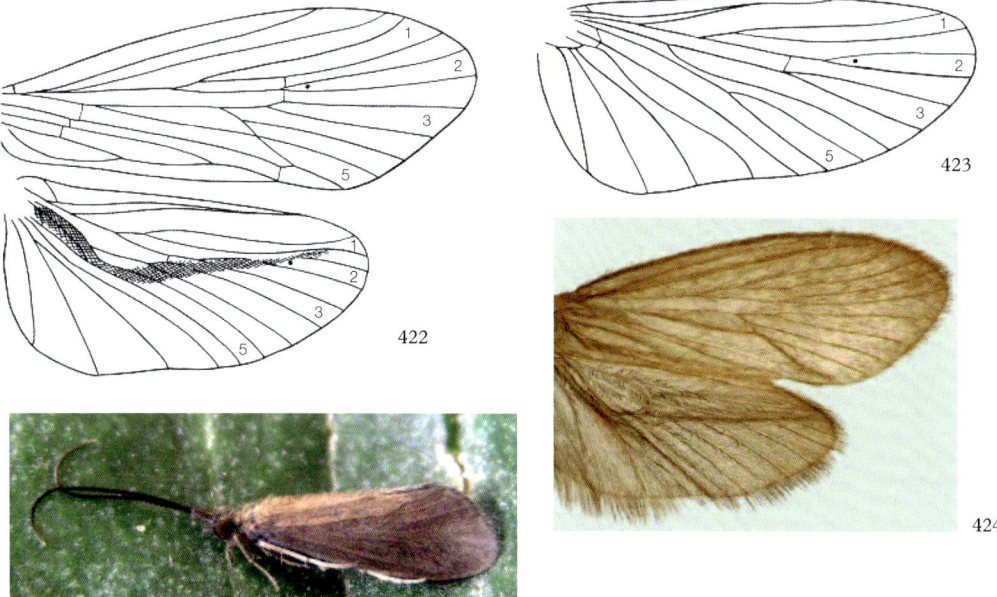

105

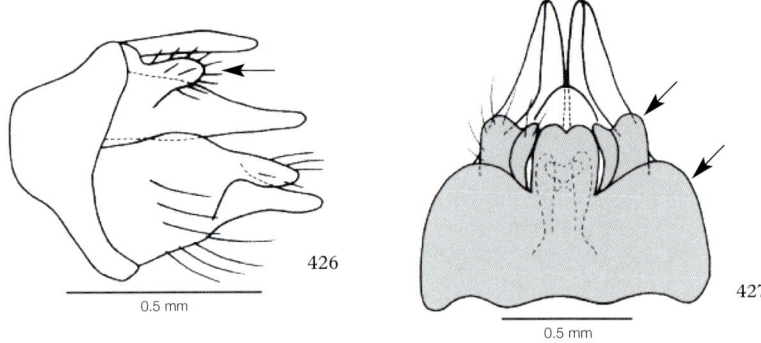

0.5 mm

426

0.5 mm

427

Figures 422-427. *Silo nigricornis*. 422 male wing venation; 423 female hind wing; 424 male wing pattern; 425 live specimen; 426 male genitalia lateral, aedeagus omitted; 427 female genitalia ventral, segment VIII shaded

## *Silo pallipes* (Fabricius, 1781)

Fore wing length: ♂ 6-9 mm, ♀ 6-10 mm (Figs 428-430); coloration as in *S. nigricornis*. Male scent hair groove on hind wing gently curved. Base of fork 1 in fore wing less than half length of discoidal cell (Fig. 428). Throughout Britain; present in Ireland; stony rivers and streams. Throughout most of Europe. Flight period: May-September. ♂ genitalia with long superior appendages (Fig. 431); ♀ genitalia with segment VIII longer than in *S. nigricornis*, and with truncate lobes (Fig. 432). The female usually has a short fork 1 in the hind wing (Fig. 429) unlike *S. nigricornis*.

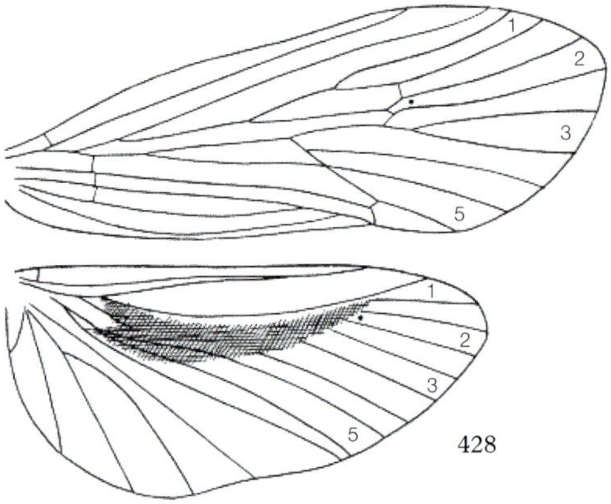

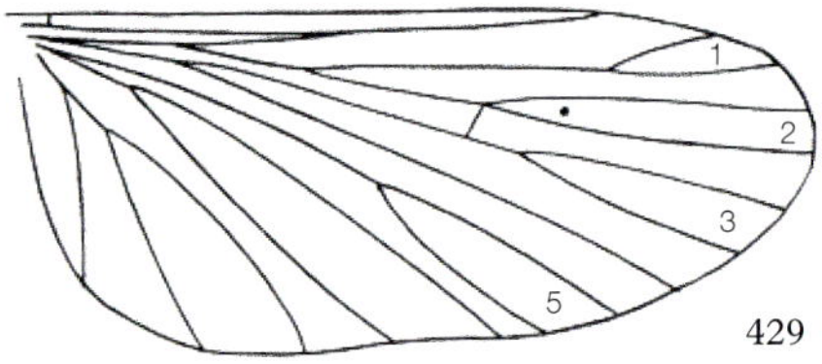

429

428

430

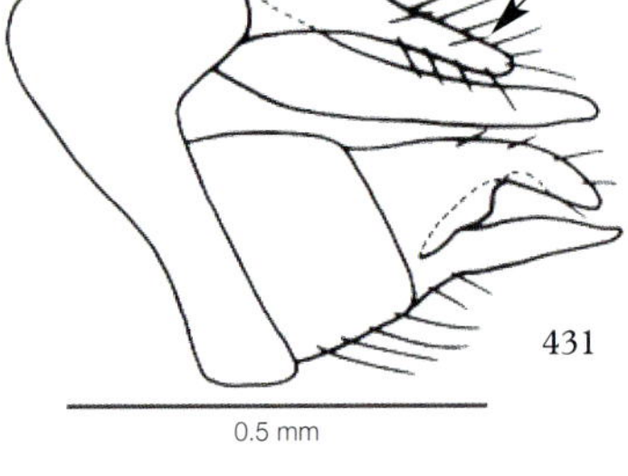

0.5 mm

431

0.5 mm

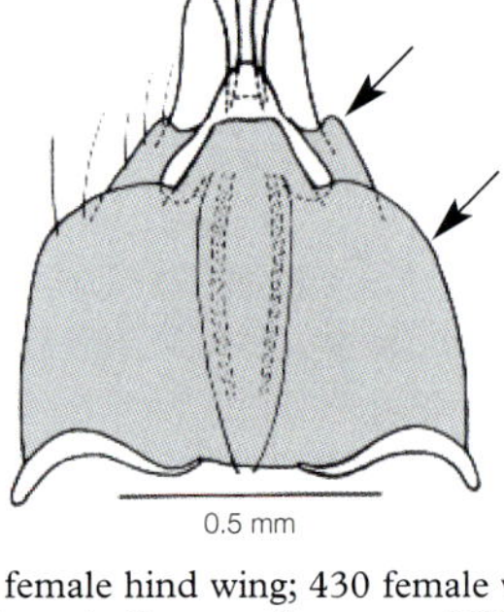

432

Figures 428-432. *Silo pallipes*. 428 male wing venation; 429 female hind wing; 430 female wing pattern; 431 male genitalia lateral, aedeagus omitted; 432 female genitalia ventral, segment VIII shaded

# Family LEPIDOSTOMATIDAE (2 genera, 3 species)

Venation described under genus below. Spur formula 2.4.4. Ocelli absent. Male maxillary palp with only three segments.

## Key to genera of Lepidostomatidae

1. In hind wing discoidal cell open (Fig. 433) ......................................... *Crunoecia* (p. 107)

- In hind wing discoidal cell closed (Fig. 434) ................................... *Lepidostoma* (p. 108)

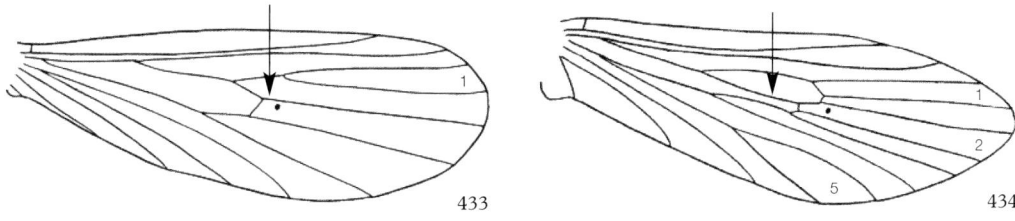

433 434

## Genus CRUNOECIA McLachlan, 1876

Fore wing with forks 1, 2 and 5 present in male, 1, 2, 3 and 5 in female; hind wing with fork 1 only in male, forks 1, 2 and 5 in female. Discoidal cell closed in fore wing, open in hind wing (Figs 435, 436). Around five species in Europe, with just one in Britain.

### *Crunoecia irrorata* (Curtis, 1834)

Fore wing length: ♂ 5-7 mm, ♀ 5-8 mm (Figs 435-437). Common throughout Britain, though with few records for SE. and E. England; present in Ireland; woodland trickles and streams. Throughout much of Europe. Flight period: May-September. ♂ genitalia as in Fig. 438; ♀ genitalia as in Fig. 439.

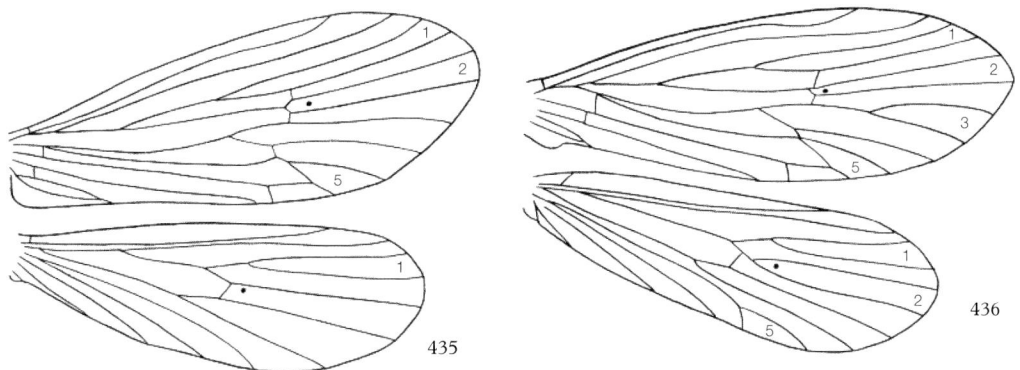

435 436

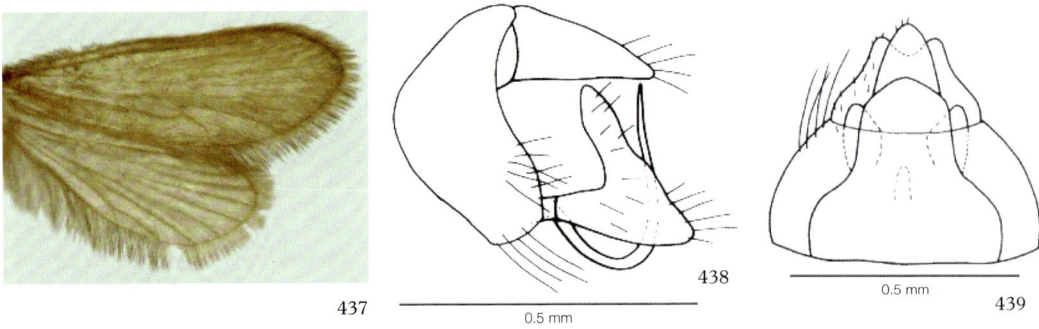

Figures 435-439. *Crunoecia irrorata*. 435 male wing venation; 436 female wing venation; 437 wing pattern; 438 male genitalia lateral, aedeagus omitted; 439 female genitalia ventral

# Genus LEPIDOSTOMA Rambur, 1842
# = LASIOCEPHALA Costa, 1857

Fore wing with forks 1 and 3 present in male, 1, 2, 3 and 5 in female; hind wing with fork 1 only in male, forks 1, 2 and 5 in female. Discoidal cell closed in both wings (Figs 440, 441). Around 10 species in Europe, of which two are found in Britain.

## *Lepidostoma hirtum* (Fabricius, 1775)
= *fimbriatum*; misidentified by some authors

*L. fimbriatum* was listed as a separate species in Mosely (1939). This is the angler's Small Silver Sedge. Fore wing length: ♂♀ 6-9 mm (Figs 440-444). Male with broad scales on fore wing Fig. 443). Common throughout Britain; present in Ireland; rivers, large streams and stony lakes. Throughout Europe. Flight period: June-September. ♂ genitalia as in Figs 445, 446; ♀ genitalia as in Fig. 447.

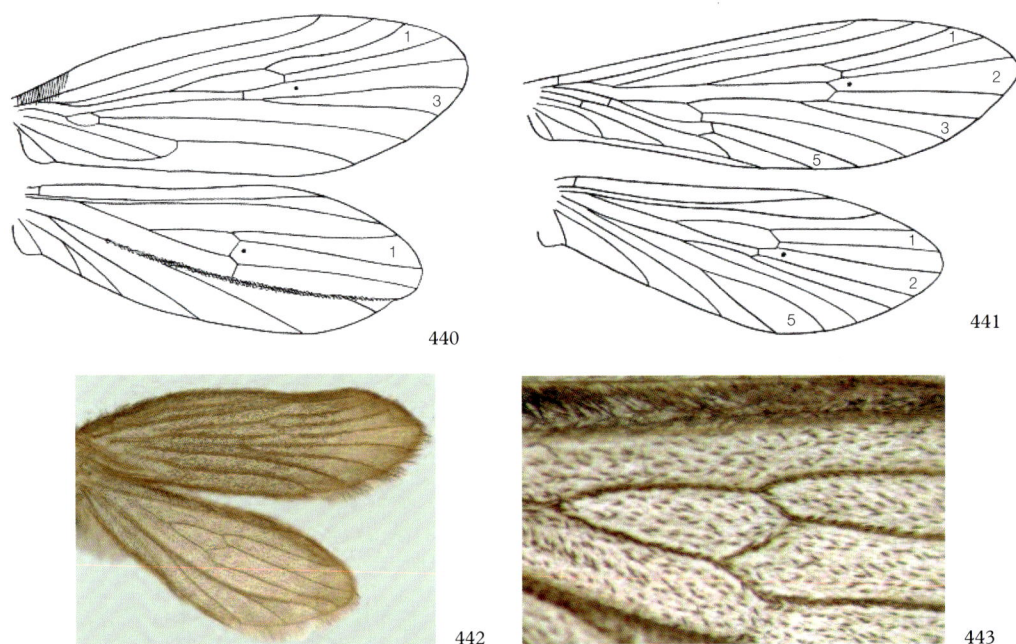

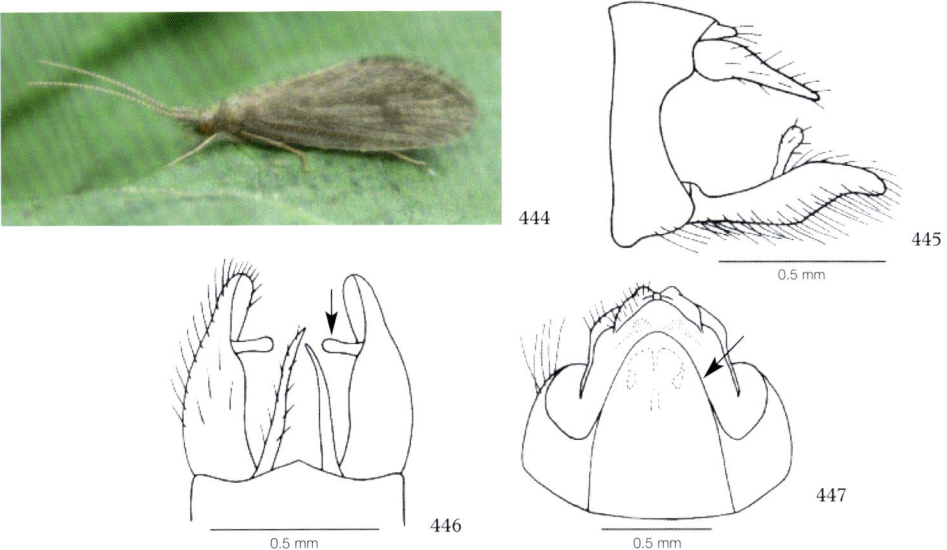

Figures 440-447. *Lepidostoma hirtum*. 440 male wing venation; 441 female wing venation; 442 male wing pattern; 443 male fore wing closeup; 444 live specimen; 445 male genitalia lateral, aedeagus omitted; 446 male genitalia ventral; 447 female genitalia ventral

## *Lepidostoma basale* (Kolenati, 1848)

Listed as *Lasiocephala basalis* in all previous British works: *Lasiocephala* is now considered a synonym of *Lepidostoma* (Weaver, 2002) so *L. basale* moves to *Lepidostoma*. Fore wing length: ♂ 6-9 mm, ♀ 8-10 mm (Figs 448-451). Male with dense fringe of hairs in a groove between costa and radius of fore wing, but no broad scales. Local throughout England, Wales and Scotland as far north as Aviemore; present in Ireland; stony rivers and large streams, larvae associated with submerged timber. Mainly southern and central Europe. Flight period: May-September. ♂ genitalia as in Figs 452, 453; ♀ genitalia as in Fig. 454.

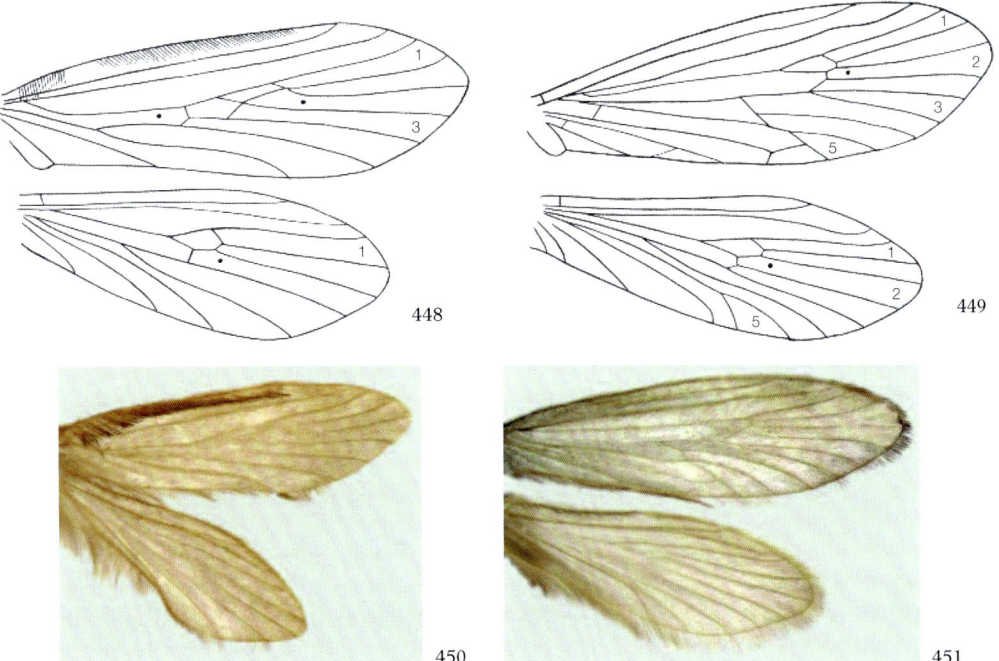

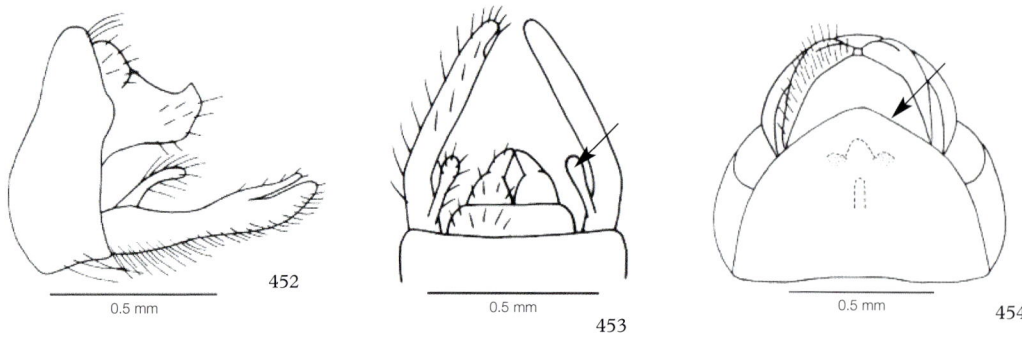

Figures 448-454. *Lepidostoma basale*. 448 male wing venation; 449 female fore wing venation; 450 male wing pattern; 451 female wing pattern; 452 male genitalia lateral, aedeagus omitted; 453 male genitalia ventral; 454 female genitalia ventral

# Family APATANIIDAE

Both wings with forks 1, 2, 3 and 5 present; discoidal cell closed in fore wing, open in hind wing (Fig. 455). Spur formula 1.2.4, ocelli present, male maxillary palp with only three segments. Until recently this group was treated as a subfamily of the Limnephilidae; it contains the single genus *Apatania*.

## Genus APATANIA Kolenati, 1848

Of around 30 European species three occur in the British Isles (one only in Ireland).

### *Apatania auricula* (Forsslund, 1930)

Listed as *Apatidea fimbriata* in Mosely (1939). *A. fimbriata* (Pictet) is a different species, found elsewhere in Europe. Fore wing length: ♂♀ 8-11 mm (Figs 455, 456); wings dark yellowish grey with distinct veins. Ireland only. Northern Europe. ♂ genitalia with short bifurcating clasper (Fig. 457); ♀ genitalia with short central vulvar scale (Fig. 458).

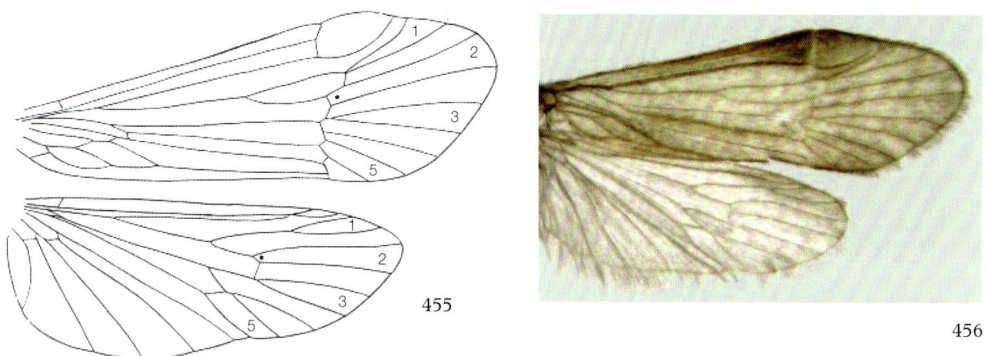

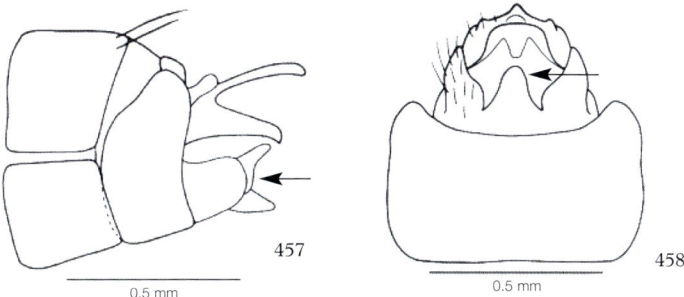

Figures 455-458. *Apatania auricula*. 455 wing venation; 456 wing pattern;
457 male genitalia lateral, aedeagus omitted; 458 female genitalia ventral

## *Apatania wallengreni* McLachlan, 1871

Fore wing length: ♂♀ 6-9 mm (Fig. 459); fore wing brown with distinct veins. Widespread in Scotland and the Lake District, and a few sites in Wales; present in Ireland; large stony lakes. Northern Europe. Flight period: April-June. Many earlier entomologists did not see this species because they missed its comparatively early flight period. ♂ genitalia with elongate two-segmented clasper (Fig. 460); ♀ genitalia with narrow central vulvar scale; lateral vulvar scales not clearly differentiated (Fig. 461).

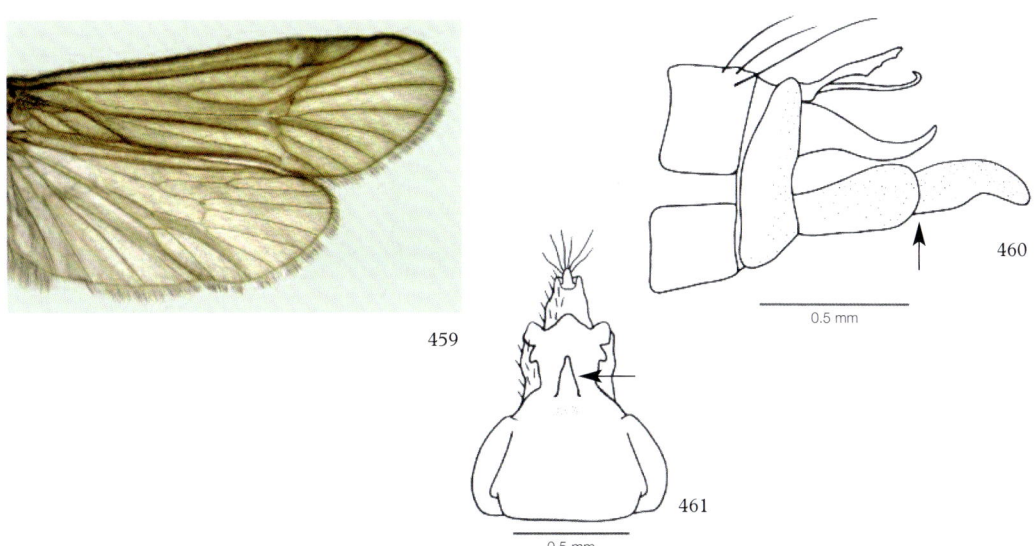

Figures 459-461. *Apatania wallengreni*. 459 wing pattern;
460 male genitalia lateral, aedeagus omitted; 461 female genitalia ventral

## *Apatania muliebris* McLachlan, 1866

Listed as *Apatidea muliebris* in Mosely (1939). This is the only parthenogenetic caddisfly in Britain and, as with many such species, the morphological differences between isolated populations can make the taxonomy difficult to interpret. The forms in Britain that resemble the Scandinavian populations of *A. nielseni* are now considered as falling within the range of *A. muliebris* (Barnard & O'Connor, 1987). Fore wing length ♀ 7-10 mm (Fig. 462); paler than the other two species of *Apatania* with less distinct veins. Scattered throughout Britain, but very local especially in lowland areas; present in Ireland; spring-fed streams. Scattered records throughout central and northern Europe. Flight period: May-June, August-October, possibly bivoltine. ♀ genitalia with very long central vulvar scale (Fig. 463).

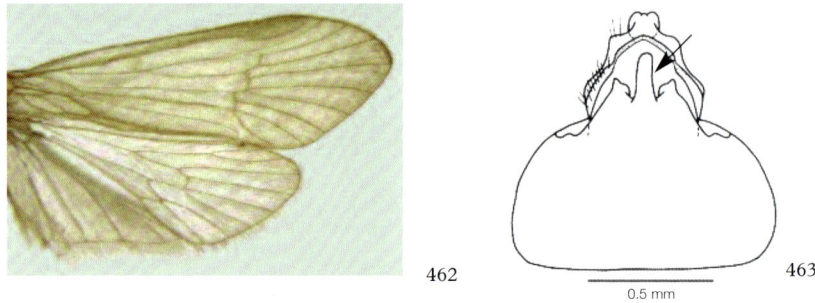

462    463

0.5 mm

Figures 462-463. *Apatania muliebris*. 462 wing pattern; 463 female genitalia ventral

# Family LIMNEPHILIDAE (19 genera, 55 species)

Both wings with forks 1, 2, 3 and 5 present; discoidal cell closed in both wings. Spurs very variable (see generic descriptions for details), ocelli present, male maxillary palp with only three segments. The classification of the subgroups within this large and complex family is still undergoing great discussion and study (e.g. Vshivkova *et al.*, 2007) and it will be some time before the relationships of all genera are clarified on a worldwide basis. The family was often spelled Limnophilidae by earlier authors, because of the early (incorrect) spelling of *Limnephilus* as *Limnophilus*.

## Key to subfamilies of the Limnephilidae

This key is not always easy to use but in doubtful cases it is a simple matter to quickly check the descriptions and illustrations of *Drusus*, *Ecclisopteryx* and *Ironoquia*, because the great majority of British species are in the Limnephilinae.

1. In hind wing forking points M1+2, M3+4 and Cu1-2 are close together, and an imaginary line joining these three points is angled (Fig. 464, circled) .................... Drusinae (p. 114)

- In hind wing these three forking points are separated, and an imaginary line joining them is almost straight (Fig. 465, circled) ........................................................................ 2

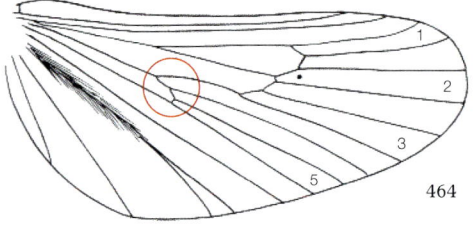

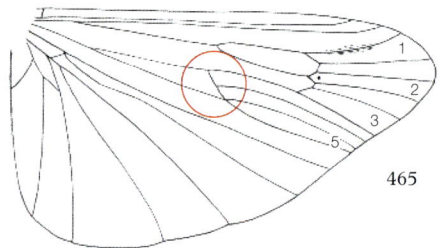

2. In fore wing base of F1 forms at least quarter of front margin of discoidal cell (Fig. 466) ..................................................................... Dicosmoecinae [*Ironoquia*] (p. 113)

- In fore wing base of F1 overlaps discoidal cell by only small amount, much less than quarter length of cell (Fig. 467) .................................................. Limnephilinae (p. 116)

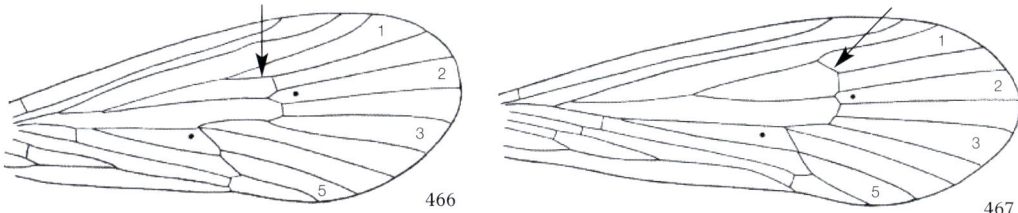

# Subfamily DICOSMOECINAE

## Genus IRONOQUIA Banks, 1916

Spur formula 1.3.4. *Ironoquia dubia* is the only European species in this genus.

### *Ironoquia dubia* (Stephens, 1837)

Listed as *Caborius dubius* in Mosely (1939). This is a BAP species: the Scarce Brown Sedge. Fore wing length: ♂ 10-11 mm, ♀ 12-13 mm (Figs 468, 469). The only recent records are from Berkshire and Hampshire; no records from Ireland; small streams. Mainly central and northern Europe. Flight period: September-October. ♂ genitalia as in Fig. 470; ♀ genitalia as in Fig. 471.

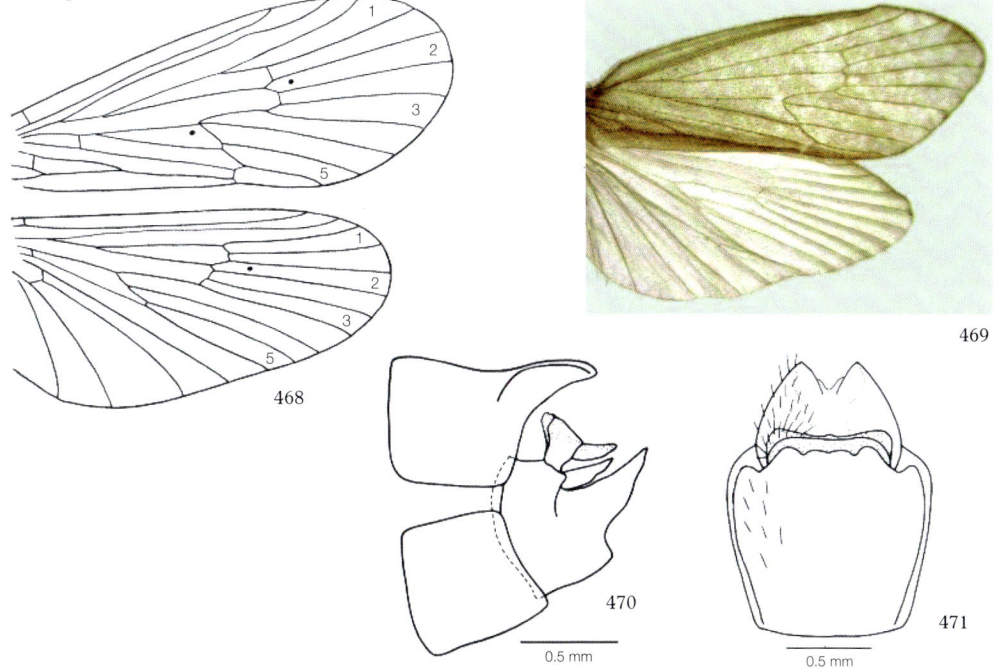

Figures 468-471. *Ironoquia dubia*. 468 wing venation; 469 wing pattern; 470 male genitalia lateral, aedeagus omitted; 471 female genitalia ventral

113

# Subfamily DRUSINAE

## Key to genera of Drusinae

1. Spur formula 1.3.3 (occasionally 0.3.3 in males) ..................................... *Drusus* (p. 114)

- Spur formula 1.2.3 .................................................................... *Ecclisopteryx* (p. 115)

# Genus DRUSUS Stephens, 1837

Spur formula 1.3.3 or 0.3.3 in males, 1.3.3 in females. There are over 50 species of *Drusus* in Europe, mainly in alpine habitats, but just one species in Britain.

## *Drusus annulatus* (Stephens, 1837)

Fore wing length: ♂ 8-13 mm, ♀ 8-14 mm (Figs 472-474). The spur formula is normally 1.3.3, but is occasionally 0.3.3 in the male. Very common throughout Britain; present in Ireland; stony streams and rivers. Mainly central Europe. Flight period: (April) May-October, possibly multivoltine (Gower, 1965). ♂ genitalia as in Fig. 475; ♀ genitalia as in Fig. 476.

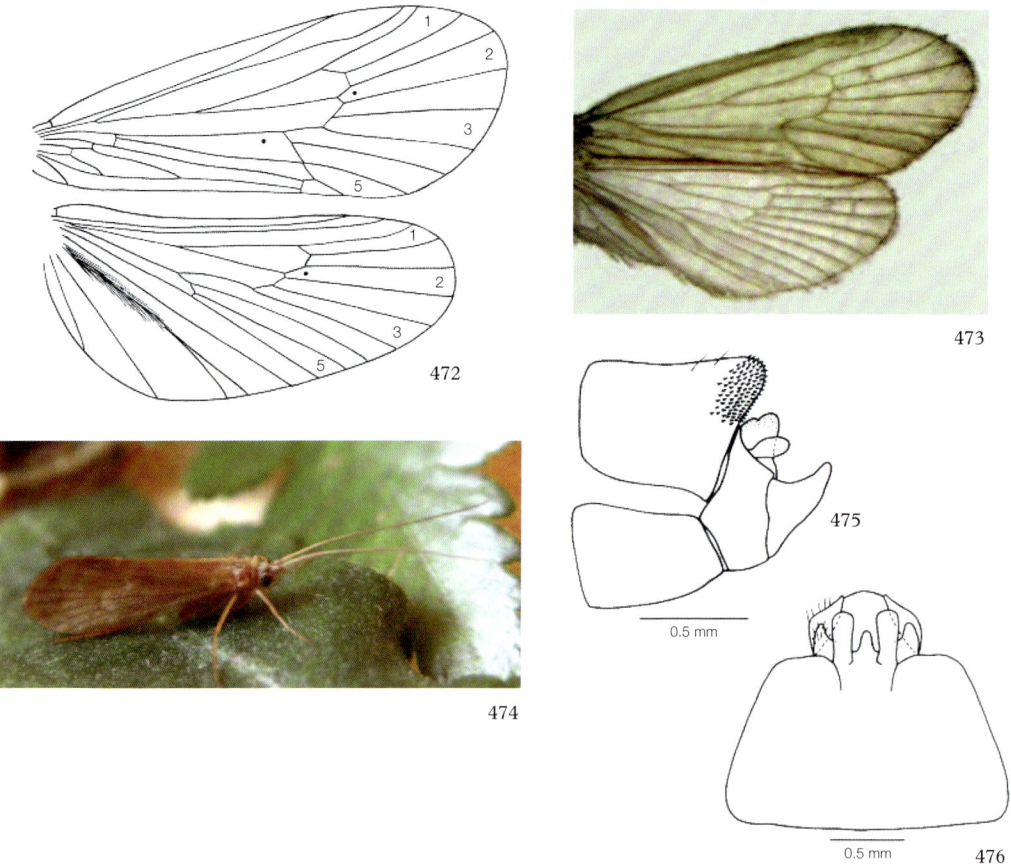

Figures 472-476. *Drusus annulatus.* 472 wing venation; 473 wing pattern; 474 live specimen; 475 male genitalia lateral, aedeagus omitted; 476 female genitalia ventral

# Genus ECCLISOPTERYX Kolenati, 1848

Spur formula 1.2.3. Only one of the five European species occurs in Britain.

## *Ecclisopteryx dalecarlica* **Kolenati, 1848**

Listed as *E. guttulata* in Mosely (1939) and Macan (1973) and all other British works. *E. dalecarlica* was originally considered as a subspecies of *E. guttulata*, but its elevation to a separate species, is confirmed by larval characters. Re-examination of British specimens, including material from Ireland (J.P. O'Connor, pers. comm.) has confirmed that only *E. dalecarlica* occurs in the British Isles. Fore wing length: ♂ 7-12 mm, ♀ 10-14 mm (Figs 477, 478). Fairly common in SW. England, Wales and the bordering English counties, N. England and Scotland; present in Ireland; fast-flowing larger streams and rivers. Mainly central and northern Europe, but the previous confusion between *E. dalecarlica* and *E. guttulata* means that the European distribution of these two species is not entirely clear. *E. guttulata* is probably the predominant species in southern Europe. Flight period: May-June (July). ♂ genitalia as in Fig. 479; ♀ genitalia as in Fig. 480.

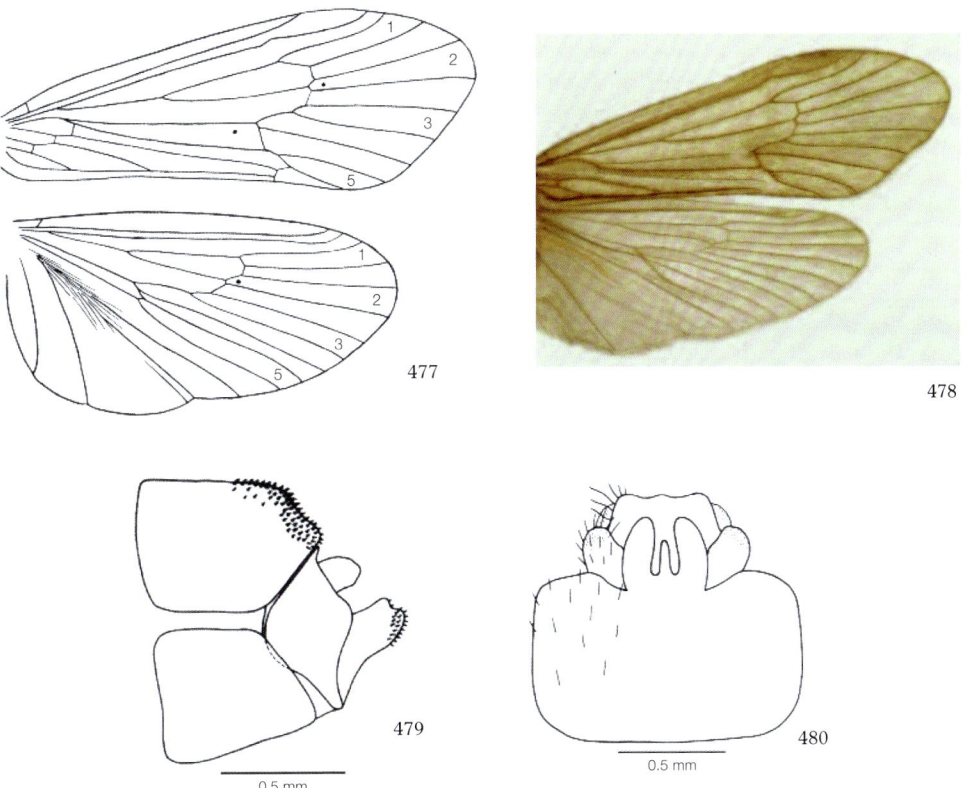

Figures 477-480. *Ecclisopteryx dalecarlica.* 477 wing venation; 478 wing pattern; 479 male genitalia lateral, aedeagus omitted; 480 female genitalia ventral

# Subfamily LIMNEPHILINAE

## Key to genera of the subfamily Limnephilinae

The close similarity between many genera means that this key is not entirely satisfactory, but it should help to place a species in approximately the right place in the family. Many species in this subfamily have distinctive wing markings which, combined with the genitalia, should make most identifications reasonably straightforward. It is worth noting that these markings are normally on the wing membrane, rather than the covering of hairs, so that even badly worn specimens can often be identified.

1. In hind wing vein R5 [RS4] marked by a dark streak (Fig. 481) .... *Grammotaulius* (p. 122)

- In hind wing vein R5 not darkly marked ................................................................. 2

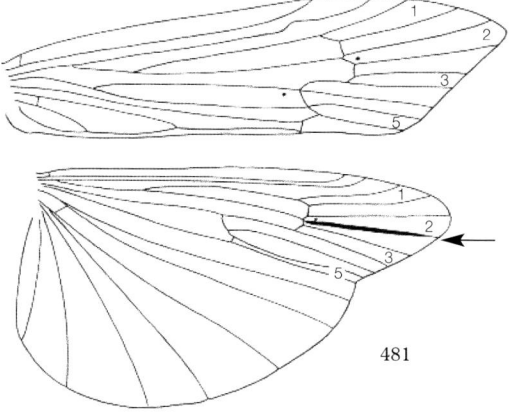

481

2. Fore wing strongly excised below apex (Fig. 482) ........................................................ 3

- Fore wing with apex straight or convex ................................................................. 4

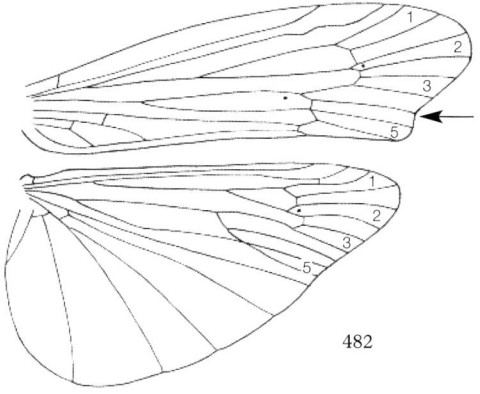

482

3. Fore wing length up to 17 mm ........................................................ *Glyphotaelius* (p. 121)

- Fore wing length 20-25 mm ........................................................ *Nemotaulius* (p. 123)

4. Margin of hind wing strongly excised behind apex, at fork 5 (Fig. 483) ...........................
.................................................................................. *Limnephilus incisus* (p. 125)

- Margin of hind wing either evenly rounded or weakly concave at fork 5 (Fig. 484) ........ 5

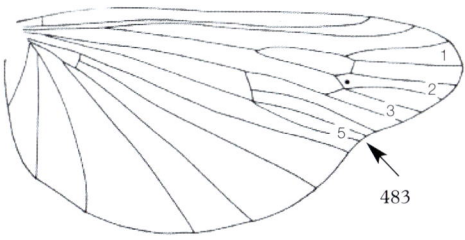
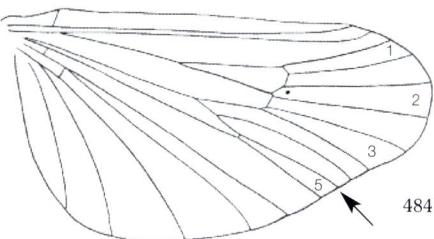

5. Fore wing membrane granular, with long bristly hairs arising from warty bases ............ 6

- Fore wing membrane smooth, hairs not noticeably long or bristly ................................ 7

6. Margin of fore wing with definite angle between forks 1 and 2 (Fig. 485) ........................
.................................................................................. *Anabolia brevipennis* (p. 120)

- Margin of fore wing evenly rounded (Fig. 486) ..................... *Chaetopteryx villosa* (p. 140)

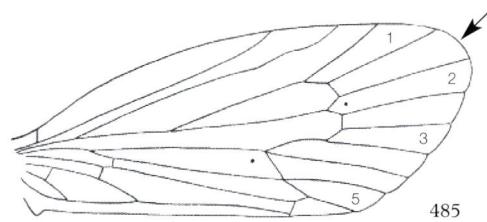
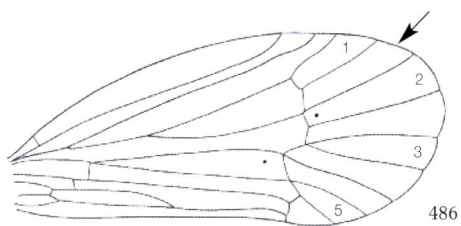

7. Fore wing long and narrow; front (costal) margin fairly straight; apical margin straight and obliquely angled. In hind wing crossvein linking F2 and F3 much shorter than one linking F1 and F2 (Fig. 487, circled). (Spur formula always 1.3.4) ............... *Limnephilus* (p. 124)

- Fore wings broader, with front (costal) margin more rounded and apex usually evenly rounded, not truncate. In hind wing crossvein linking F3 and F2 usually as long as or longer than one linking F1 and F2 (Fig. 488, circled). (Spurs variable) ........................ 8

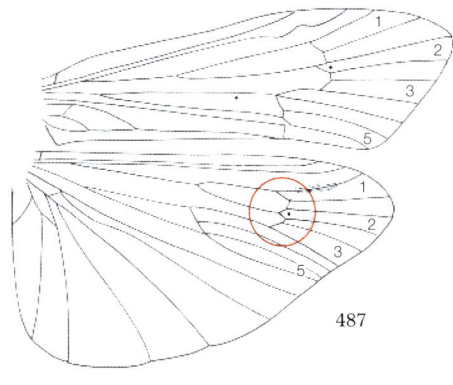
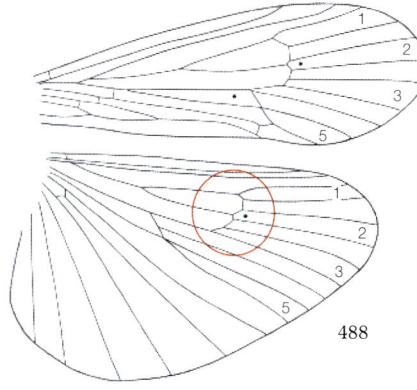

8. Hind tibia with 2 or 3 spurs, spur formula 0.2.2, or 1.3.3 ............................................ 15

- Hind tibia with 4 spurs, spur formula 0.3.4 or 1.3.4 .................................................... 9

9. Vein R2 (RS1) in fore wing strongly concave where it forms front margin of discoidal cell (Fig. 489) ..................................................................................................................... 10

- Vein R2 (RS1) in fore wing straight or only slightly curved (Fig. 490) ......................... 12

10. Fore wing length usually greater than 18 mm ................................. *Stenophylax* (p. 143)

- Fore wing length usually less than 18 mm .................................................................. 11

11. Fore wing greyish brown ............................................................... *Mesophylax* (p. 141)

- Fore wing yellowish brown ........................................................... *Micropterna* (p. 142)

12. Coloration of fore wing uniformly dark-brown, reddish or yellow-brown, without any distinctive markings ...................................................................................................... 13

- Fore wing with clear white or translucent markings, usually forming streaks or points within cells ............................................................................................................... 14

13. Fore wing dark reddish-brown in life, often with two pale marks at thyridium and arculus ........................................................................................................... *Anabolia* (p. 119)

- Fore wing shining blackish brown, with very dark veins ............... *Hydatophylax* (p. 149)

14. Large species: fore wing length 15-25 mm. Fore wing strongly marked, with pale longitudinal streaks within cells ................................................... *Potamophylax* (p. 144)

- Smaller species: fore wing length up to 14 mm. Fore wings grey or greyish brown, with whitish patches ........................................................................ *Rhadicoleptus* (p. 124)

15. Spur formula 1.3.3 ...................................................................................................... 16

- Spur formula 0.2.2. (R1 in male fore wing linked by short crossvein with subcosta; female wingless) ........................................................................... *Enoicyla* (p. 148)

16. Fore wing 17-25 mm; yellow with dark streaks in most cells ................. *Halesus* (p. 147)

- Fore wing up to 14 mm; no darker streaks in cells ..................................................... 17

17. Fore wing smoky black, with dark pterostigma and pale thyridial spot ... *Allogamus* (p. 146)

- Fore wing yellowish brown; pterostigma only slightly darker, and thyridial spot not very pronounced ......................................................................... *Melampophylax* (p. 148)

# Tribe Limnephilini

There have been many attempts to clarify the generic limits in this tribe, with various synonymies proposed (e.g. Grigorenko, 2002) but these have not received general acceptance. Far more work is needed on these genera on a worldwide revisionary scale before any meaningful conclusions can be reached (see also Vshivkova *et al.*, 2007).

## Genus ANABOLIA Stephens, 1837
## = PHACOPTERYX Kolenati, 1848

Spur formula 1.3.4. Of the eight European species two are found in Britain.

### *Anabolia nervosa* (Curtis, 1834)

This is the angler's Brown Sedge. Fore wing length: ♂ 10-14 mm, ♀ 10-15 mm (Figs 491-493); two distinct pale spots near the hind margin of fore wing. Common throughout Britain; present in Ireland; large streams, rivers, lakes and large ponds. Throughout much of Europe except the east. Flight period: (July) August-October. ♂ genitalia as in Fig. 494; ♀ genitalia as in Fig. 495.

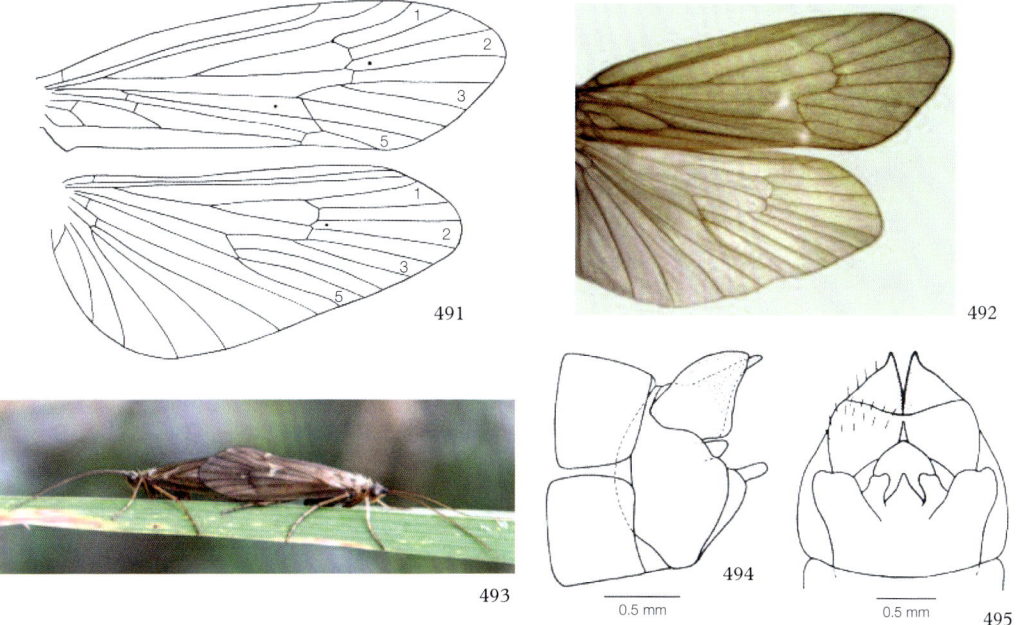

Figures 491-495. *Anabolia nervosa*. 491 wing venation; 492 wing pattern; 493 mating pair;
494 male genitalia lateral, aedeagus omitted; 495 female genitalia ventral

119

## *Anabolia brevipennis* (Curtis, 1834)

Listed as *Phacopteryx brevipennis* in Mosely (1939) and other works. Fore wing length: ♂♀ 8-10 mm (Figs 496, 497) fore wing membrane with granular appearance and long bristly hairs. Very local in E. Anglia, NW. England, Yorkshire, and single records from Wales and Dumfriesshire; present in Ireland; seasonal pools in carr woodland and old *Phragmites* fens. Mainly central and northern Europe. Flight period: August-October, with summer diapause. ♂ genitalia as in Fig. 498; ♀ genitalia as in Fig. 499.

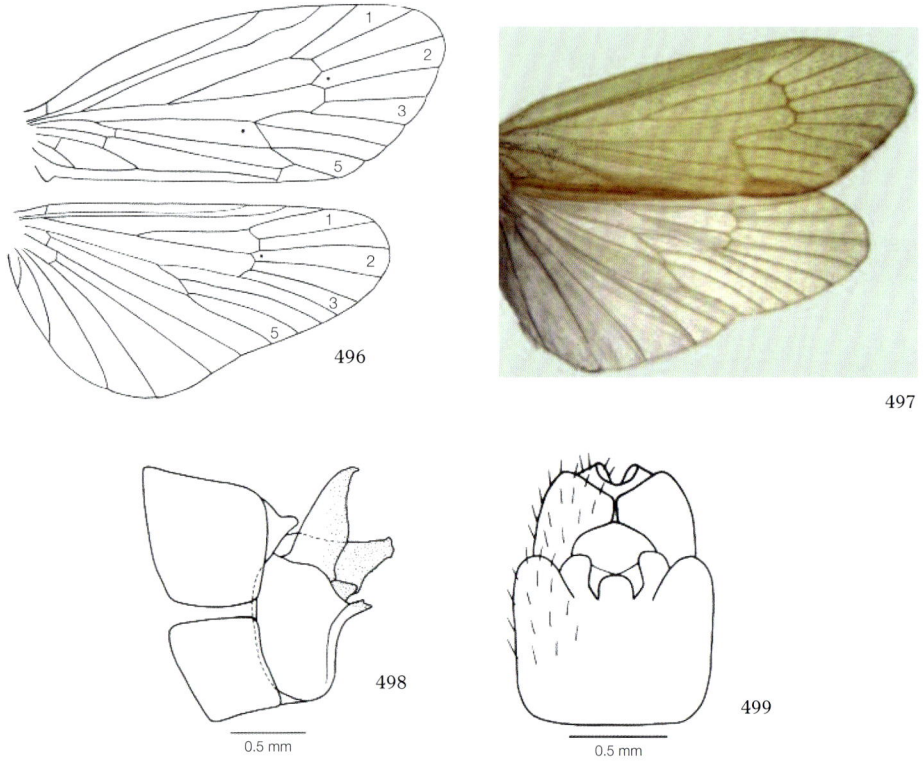

Figures 496-499. *Anabolia brevipennis*. 496 wing venation; 497 wing pattern; 498 male genitalia lateral, aedeagus omitted; 499 female genitalia ventral

# Genus GLYPHOTAELIUS Stephens, 1837

Spur formula 1.3.4. *G. pellucidus* is the only British species; two more occur in Europe.

## *Glyphotaelius pellucidus* (Retzius, 1783)

This is the angler's Mottled Sedge. Fore wing length: ♂♀ 12-17 mm (Figs 500-503): one of the few species with marked sexual dimorphism in wing pattern. This is the only common species with a strongly excised outer margin in the fore wing (Fig. 500). Common throughout Britain, though less common in Scotland; present in Ireland; temporary pools and streams, often in woodland. Throughout Europe. Flight period: May-June, August-October, with summer diapause. ♂ genitalia as in Fig. 504; ♀ genitalia as in Fig. 505.

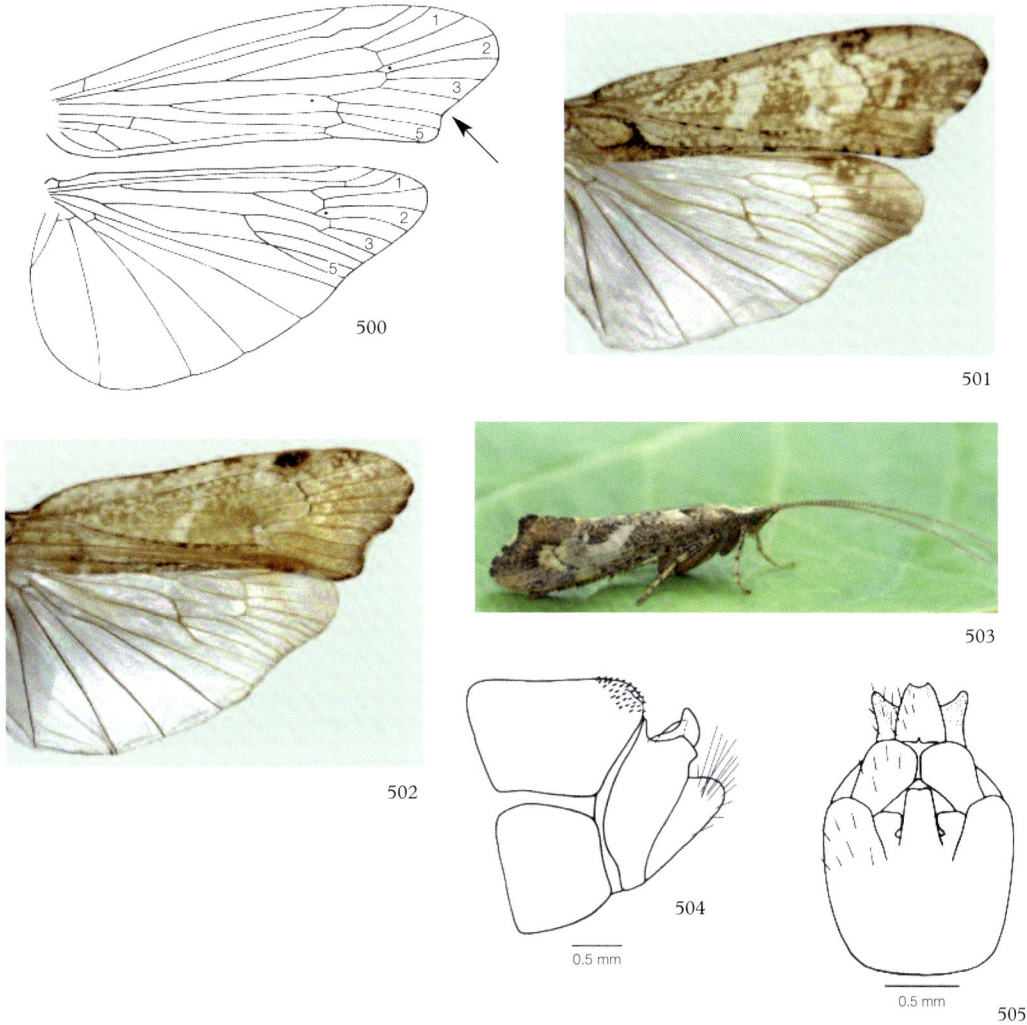

Figures 500-505. *Glyphotaelius pellucidus*. 500 wing venation; 501 male wing pattern; 502 female wing pattern; 503 live specimen; 504 male genitalia lateral, aedeagus omitted; 505 female genitalia ventral

# Genus GRAMMOTAULIUS Kolenati, 1848

Spur formula 1.3.4. Five species in Europe, with two in Britain. In both species the dark line on the hind wing shows through the fore wing when at rest.

## *Grammotaulius nigropunctatus* (Retzius, 1783)
= *atomarius* (Fabricius, 1793)

Listed as *G. atomarius* in Mosely (1939) and Macan (1973) Fore wing length: ♂♀ 14-20 mm (Figs 506-508); fore wing heavily spotted with black. Fairly common throughout Britain; present in Ireland; seasonal ponds, marshes and ditches that dry up in the summer. Throughout Europe. Flight period: May-July, August-October, with summer diapause. ♂ genitalia with broad U-shaped excision in apex of superior appendages, intermediate appendages relatively narrow and pointed (Fig. 509); ♀ genitalia with central vulvar scale longer than lateral scales (Fig. 510).

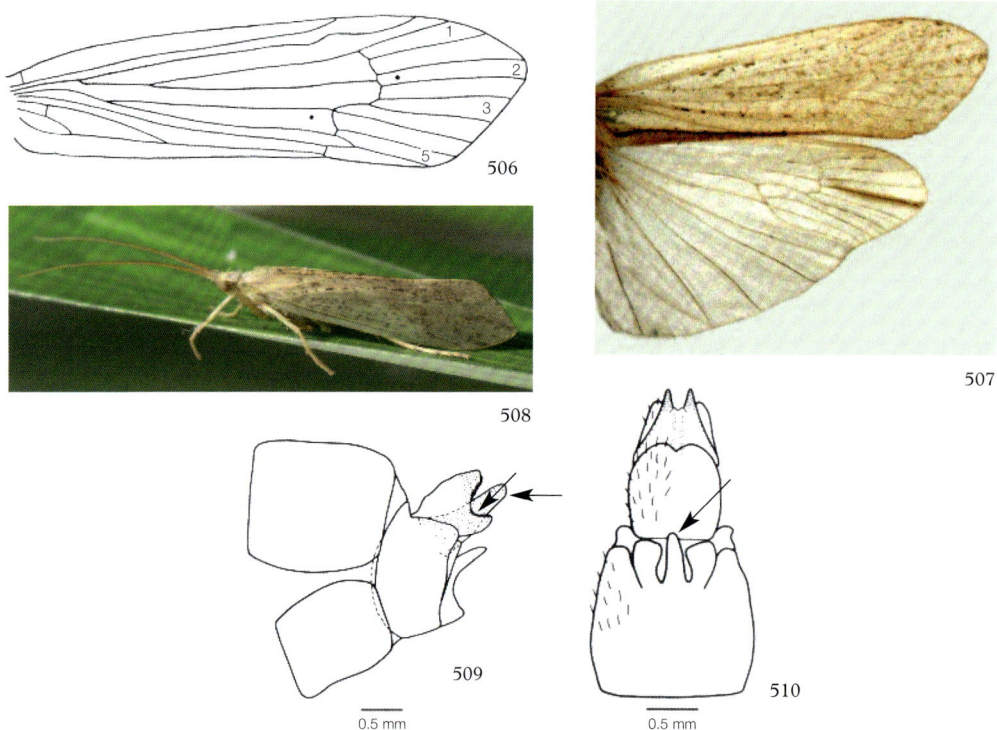

Figures 506-510. *Grammotaulius nigropunctatus.* 506 wing venation; 507 wing pattern; 508 live specimen; 509 male genitalia lateral, aedeagus omitted; 510 female genitalia ventral

## *Grammotaulius nitidus* (Müller, 1764)

Fore wing length: ♂ 17-23 mm, ♀ 17-24 mm (Figs 511-513): fore wing more acutely pointed than in *G. nigropunctatus*, but paler and unicolorous. Very local in East Anglia (Ross, 2006), Somerset and north Kent; no records from Ireland; apparently associated with sites that have extensive reed swamp. Throughout much of Europe. Flight period: July-September. ♂ ♂ genitalia with small V-shaped excision in apex of superior appendages, intermediate appendages broader and blunter than in *G. nigropunctatus* (Fig. 514); ♀ genitalia with central vulvar scale about same length as lateral scales (Fig. 515).

511

512

513

514

515

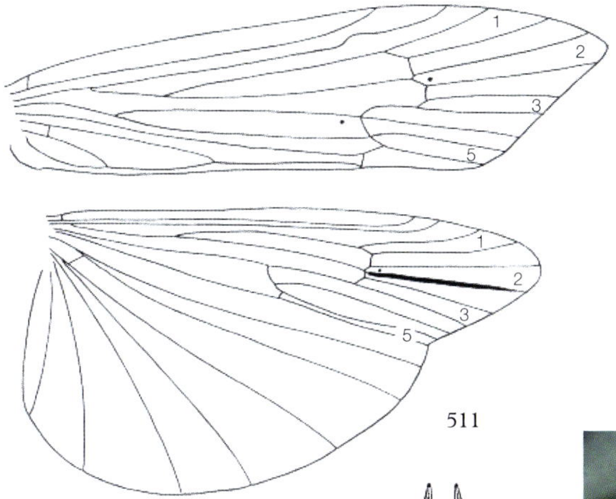

Figures 511-515. *Grammotaulius nitidus.* 511 wing venation; 512 wing pattern; 513 live specimen; 514 male genitalia lateral, aedeagus omitted; 515 female genitalia ventral

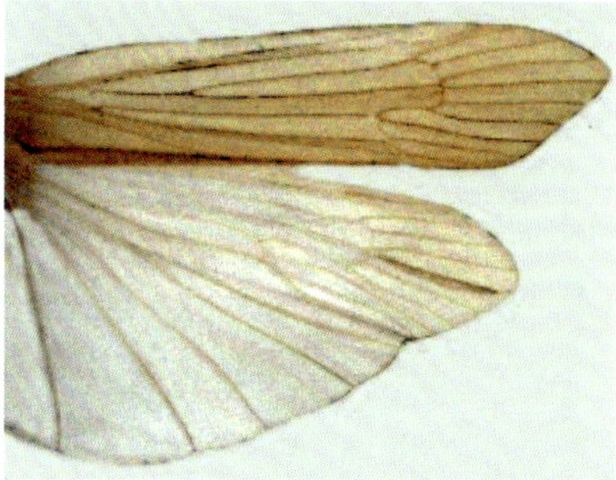

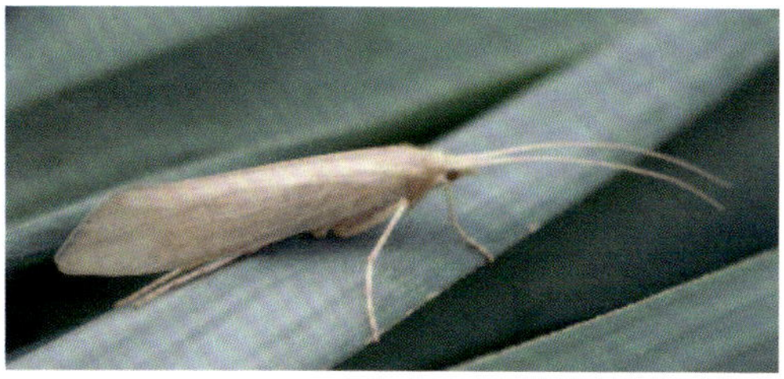

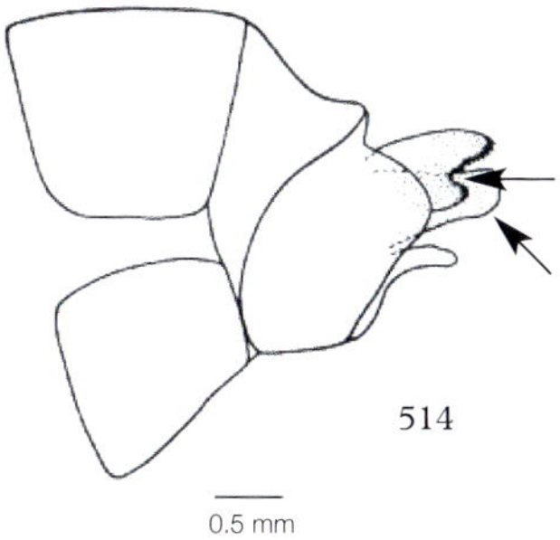

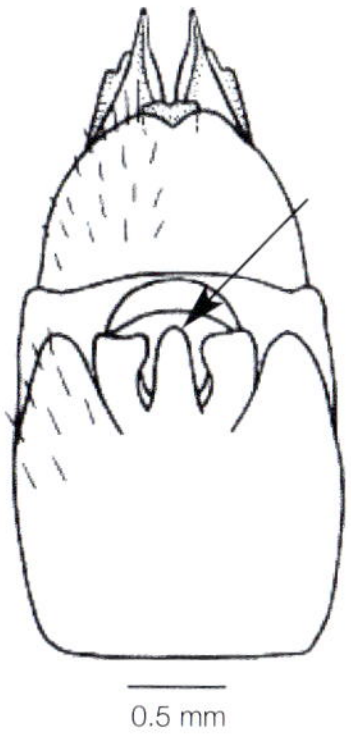

## Genus NEMOTAULIUS Banks, 1906

Spur formula 1.3.4. *N. punctatolineatus* is the only European species.

### *Nemotaulius punctatolineatus* (Retzius, 1783)

First found in Britain by Pelham-Clinton (1966a). Fore wing length: ♂♀ 22-26 mm (Fig. 516), venation similar to *Glyphotaelius*. The wing shape is also similar to that of *Glyphotaelius*, but this is a much larger species. May be locally common in Caithness and Sutherland, also in Aviemore area; no records from Ireland; pools in blanket bog. Mainly western and northern Europe. Flight period: June-July. ♂ genitalia as in Fig. 517; ♀ genitalia as in Fig. 518.

516

517

518

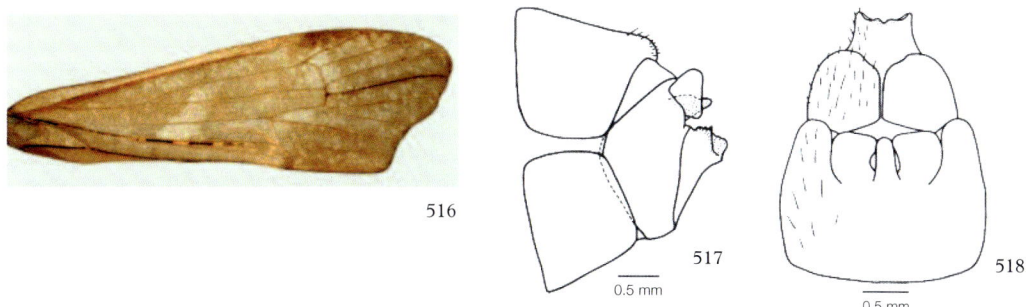

Figures 516-518. *Nemotaulius punctatolineatus.* 516 fore wing pattern; 517 male genitalia lateral, aedeagus omitted; 518 female genitalia ventral

# Genus RHADICOLEPTUS Wallengren, 1891

Spur formula 1.3.4. One species in Britain, with one other species in Europe.

### Rhadicoleptus alpestris (Kolenati, 1848)

Listed as *Stenophylax alpestris* in Mosely (1939). Fore wing length: ♂ 10-14 mm, ♀ 11-14 mm (Fig. 519); fore wing greyish yellow with large white patches in life. Local in Wales and N. England, with some records from N. Midlands, southern half of Scotland and Aviemore area; no records from Ireland; raised bog pools and upland watershed mires. Throughout much of Europe. Flight period: May-August. ♂ genitalia as in Fig. 520; ♀ genitalia as in Fig. 521.

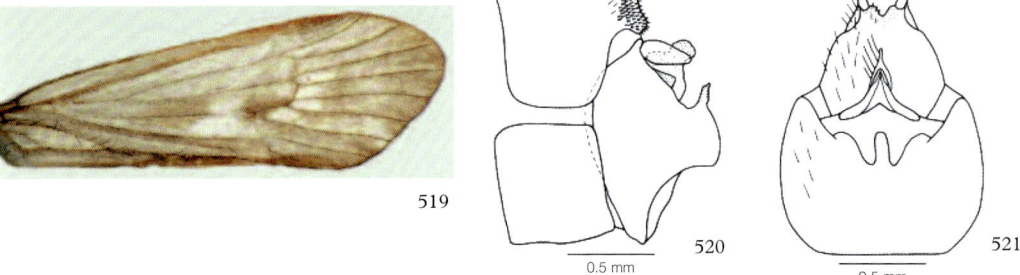

Figures 519-521. *Rhadicoleptus alpestris.* 519 fore wing pattern; 520 male genitalia lateral, aedeagus omitted; 521 female genitalia ventral

# Genus LIMNEPHILUS Leach, 1815

Spur formula 1.3.4. Of over 60 species in Europe there are 29 in the British Isles (one found only in Ireland), making this the largest British caddisfly genus. There have been some earlier attempts at conventional keys to the British species of *Limnephilus*, e.g. Tindall (1963), but these are never easy to use and inevitably involve comparing illustrations of genitalia, hence the pictorial approach in this handbook.

### Limnephilus affinis Curtis, 1834

Fore wing length: ♂ ♀ 9-13 mm (Figs 522, 523); fore wing strongly marked with a very dark pterostigma. Common throughout Britain; present in Ireland; still or slow water which may be temporary, and is tolerant of salinity and some pollution. Throughout Europe. Flight period: April-June, August-October, with summer diapause. ♂ genitalia as in Fig. 524; ♀ genitalia as in Fig. 525.

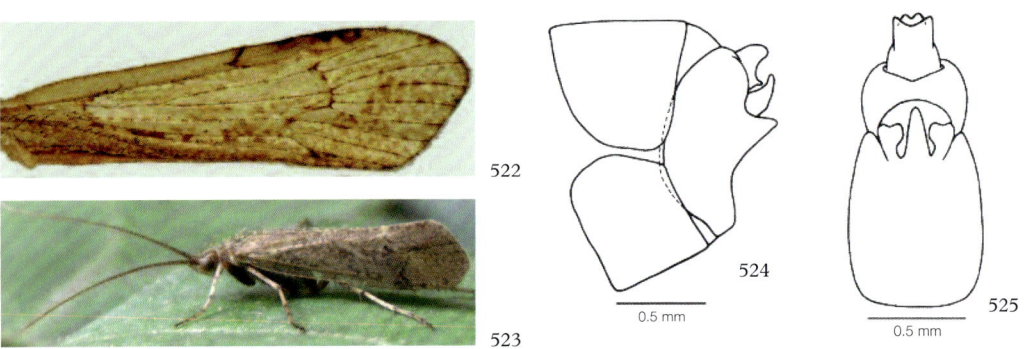

Figures 522-525. *Limnephilus affinis.* 522 fore wing pattern; 523 live specimen; 524 male genitalia lateral, aedeagus omitted; 525 female genitalia ventral

## *Limnephilus incisus* Curtis, 1834

Listed as *Colpotaulius incisus* in Mosely (1939). This is a very distinctive species that probably belongs in a separate genus, although the genitalia suggest some affinities with *L. affinis*. Fore wing length: ♂♀ 6-10 mm (Figs 526, 527): the fore wing shape, with the convex costal margin, is like no other in the genus. The front leg of the male has dark fringes of hairs along the femur and tibia, and the spur is dark and slightly hooked (Fig. 528). Common throughout Britain, but few records from S. England; present in Ireland; marshes, ditches, fens. Throughout much of Europe. Flight period: May-September, with summer diapause. ♂ genitalia as in Fig. 529; ♀ genitalia as in Fig. 530.

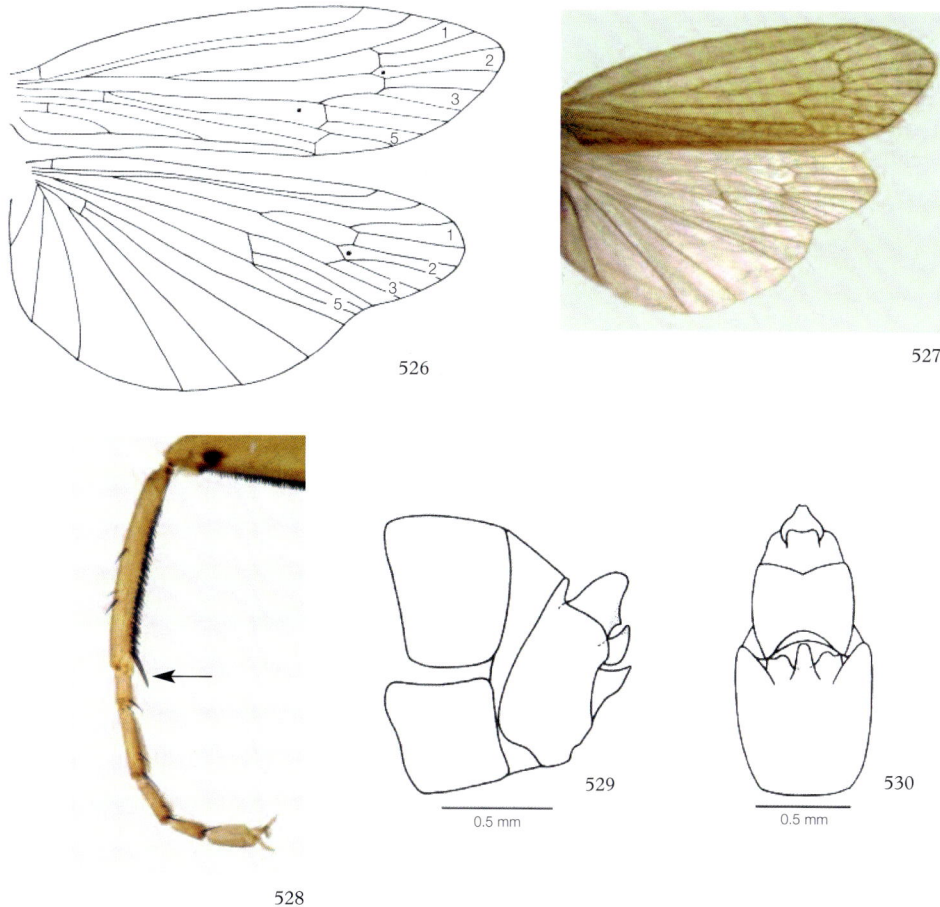

Figures 526-530. *Limnephilus incisus*. 526 wing venation; 527 wing pattern; 528 male fore leg with modified spur; 529 male genitalia lateral, aedeagus omitted; 530 female genitalia ventral

## *Limnephilus rhombicus* (Linnaeus, 1758)

Fore wing length: ♂♀ 14-19 mm (Figs 531-533); the large pale rhomboid marking near the centre of the fore wing that gives the species its name is usually diagnostic, though *L. marmoratus* may look similar. Common throughout Britain; present in Ireland; ponds, lakes, marshes and rivers. Throughout Europe. Flight period: (April) May-June, July-September, with summer diapause. ♂ genitalia as in Fig. 534; ♀ genitalia as in Fig. 535.

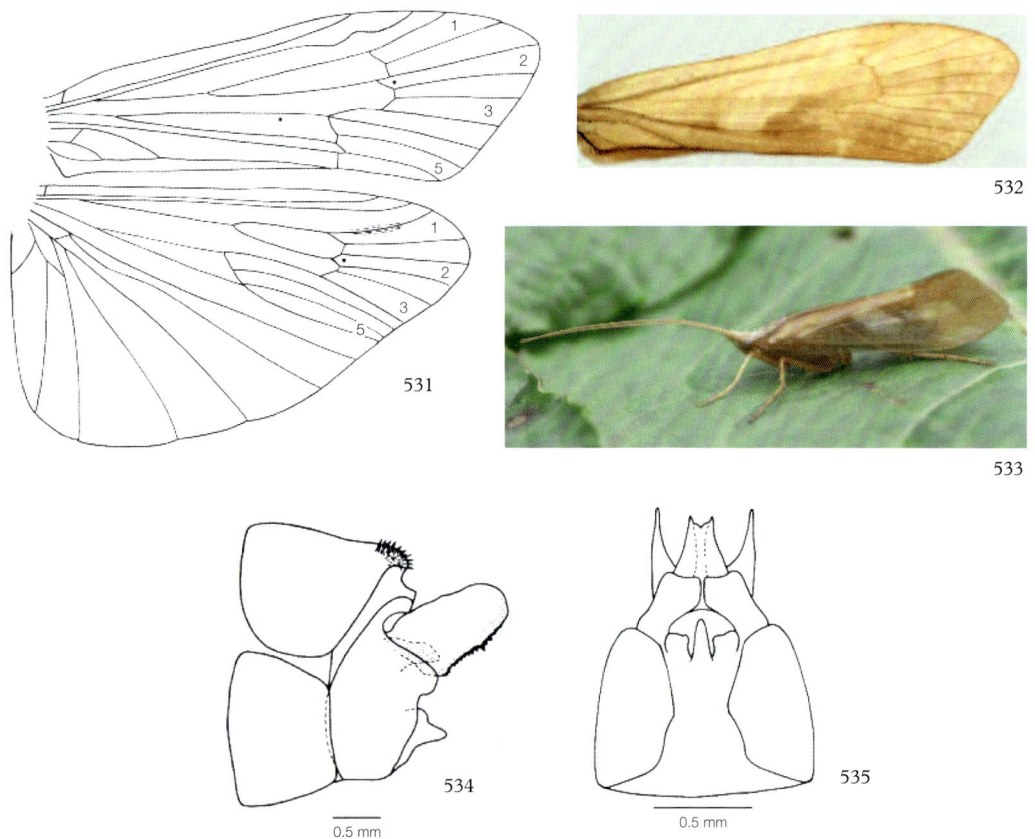

Figures 531-535. *Limnephilus rhombicus.* 531 wing venation; 532 fore wing pattern; 533 live specimen; 534 male genitalia lateral, aedeagus omitted; 535 female genitalia ventral

## *Limnephilus flavicornis* (Fabricius, 1787)

Fore wing length: ♂♀ 11-17 mm (Figs 536, 537); a very pale species often with a rather greasy appearance. Common throughout Britain except for uplands and Scotland; present in Ireland; ponds, lakes and pools (sometimes temporary). Throughout most of Europe. Flight period: May-July, August-November, with summer diapause. ♂ genitalia as in Fig. 538; ♀ genitalia as in Fig. 539.

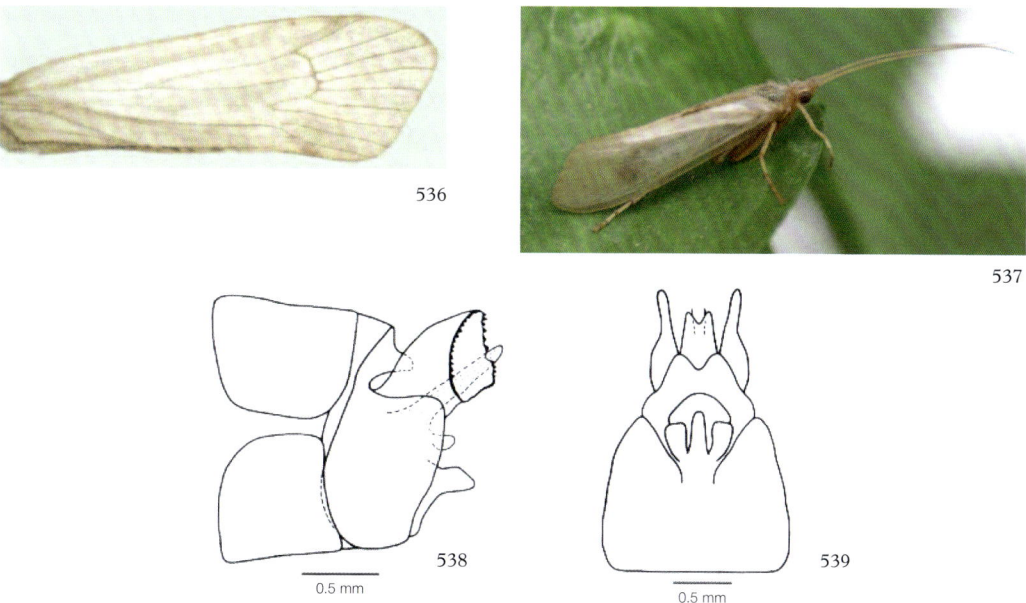

Figures 536-539. *Limnephilus flavicornis*. 536 fore wing pattern; 537 live specimen;
538 male genitalia lateral, aedeagus omitted; 539 female genitalia ventral

## *Limnephilus borealis* (Zetterstedt, 1840)

Fore wing length: ♂♀ 14-17 mm (Fig. 540): with a pale crescent-shaped marking at the apex of the fore wing like *L. lunatus*, but the marking is not bordered by a dark line. Scotland (Highland), sometimes locally common; no records from Ireland; lakes and pools. Mainly northern Europe, and the Alps. Flight period: August-September. ♂ genitalia as in Fig. 541; ♀ genitalia as in Fig. 542.

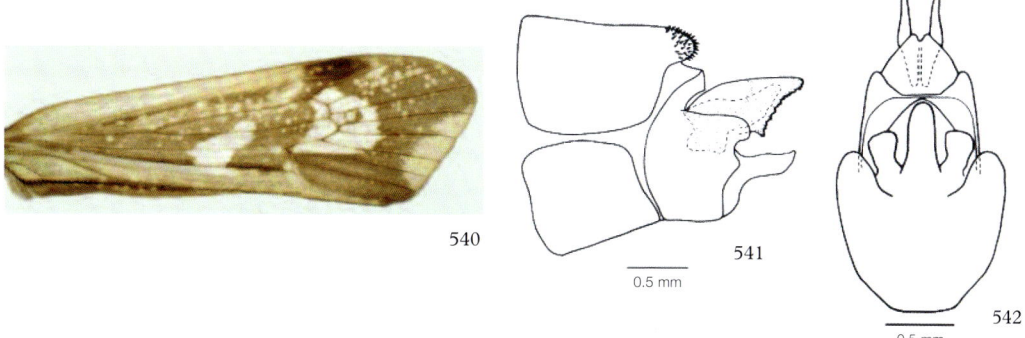

Figures 540-542. *Limnephilus borealis*. 540 fore wing pattern;
541 male genitalia lateral, aedeagus omitted; 542 female genitalia ventral

## *Limnephilus subcentralis* (Brauer, 1857)

Fore wing length: ♂♀ 11-14 mm (Fig. 543): as in *L. borealis* the apical pale crescent has no dark margin. Central Scotland, not common; no records from Ireland; lakes and ponds. Throughout much of Europe except extreme south. Flight period: May-October, apparently with a summer diapause. ♂ genitalia as in Fig. 544; ♀ genitalia as in Fig. 545.

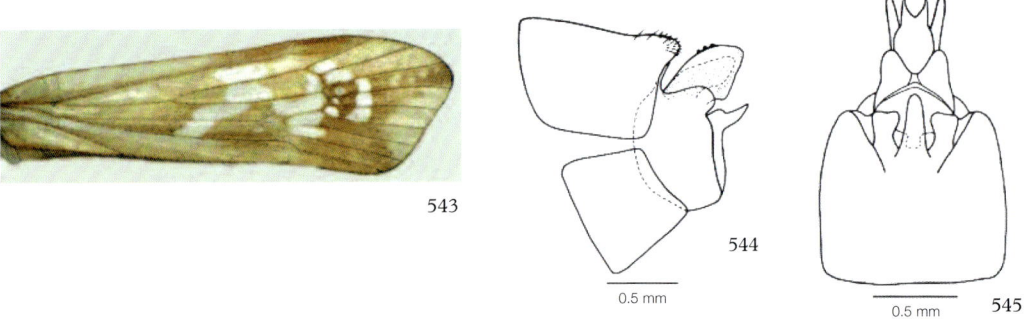

Figures 543-545. *Limnephilus subcentralis.* 543 fore wing pattern; 544 male genitalia lateral, aedeagus omitted; 545 female genitalia ventral

## *Limnephilus lunatus* Curtis, 1834

This species, and *L. marmoratus*, are the angler's Cinnamon Sedges. Fore wing length: ♂ 10-14 mm, ♀ 10-15 mm (Figs 546, 547): by far the most common species that has a pale crescent-shaped marking at the apex of the fore wing, and this marking is always bordered by a dark line. One of the commonest caddisflies, widespread throughout Britain; present in Ireland; streams, ponds, lakes and marshes. Throughout Europe. Flight period: May-November; those emerging in the spring probably have a summer diapause, but most emerge in the autumn. ♂ genitalia as in Fig. 548; ♀ genitalia as in Fig. 549.

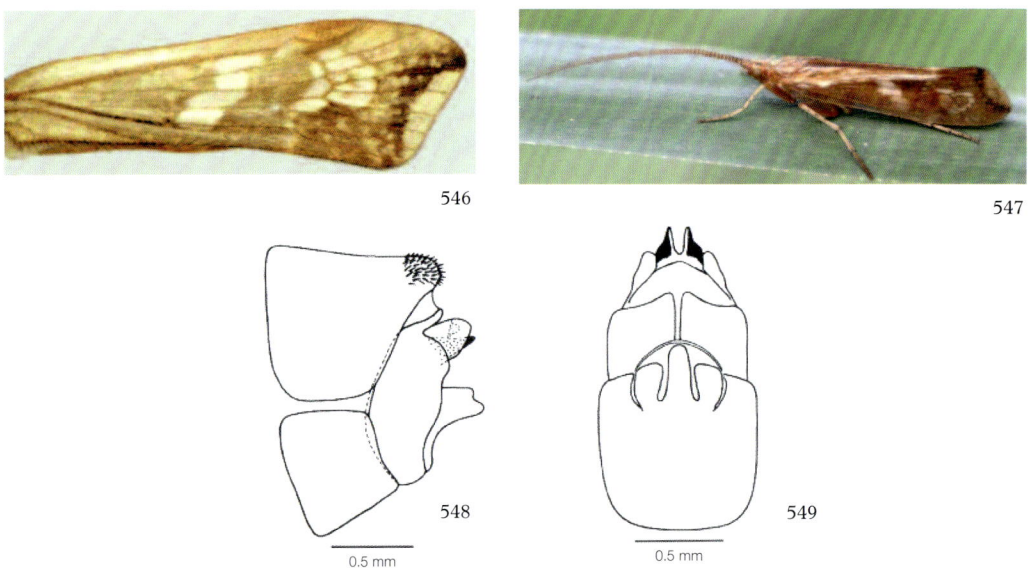

Figures 546-549. *Limnephilus lunatus.* 546 fore wing pattern; 547 live specimen; 548 male genitalia lateral, aedeagus omitted; 549 female genitalia ventral

## *Limnephilus marmoratus* Curtis, 1834

This species, and *L. lunatus*, are the angler's Cinnamon Sedges. Fore wing length: ♂ 12-16 mm, ♀ 12-17 mm (Figs 550, 551); usually more strongly marked than *L. rhombicus* though it can appear similar. Common throughout Britain; present in Ireland; ponds, lakes and pools (sometimes temporary). Throughout most of Europe, except east. Flight period: (May) June-October (November), with summer diapause. ♂ genitalia as in Fig. 552; ♀ genitalia as in Fig. 553.

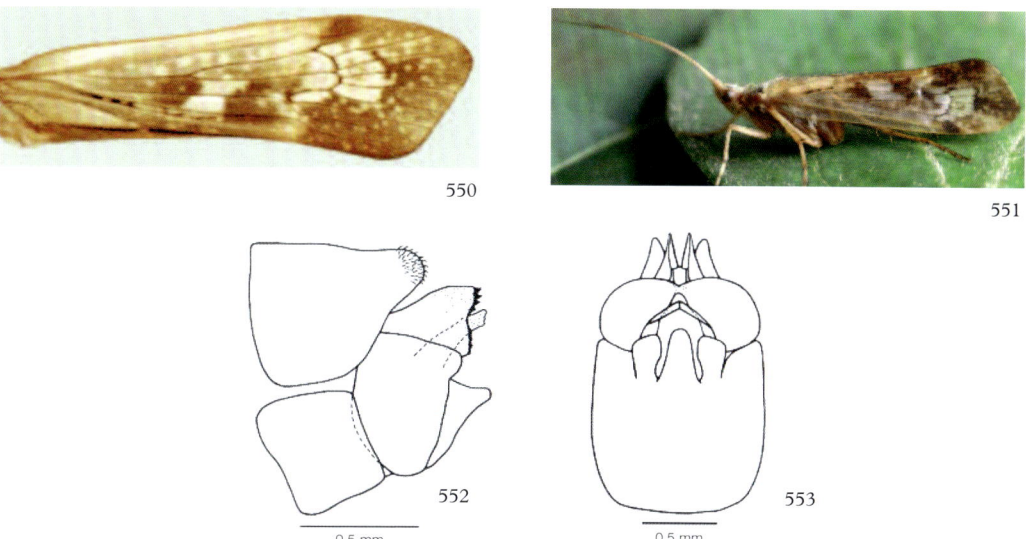

Figures 550-553. *Limnephilus marmoratus*. 550 fore wing pattern; 551 live specimen; 552 male genitalia lateral, aedeagus omitted; 553 female genitalia ventral

## *Limnephilus stigma* Curtis, 1834

Fore wing length: ♂ ♀ 13-17 mm (Fig. 554): the colour pattern can be very variable or even absent though the pterostigma is always strongly marked. Local throughout Britain, few records from S. England; present in Ireland; marshes and ponds. Throughout Europe. Flight period: June-July, August-October, with summer diapause. ♂ genitalia as in Fig. 555; ♀ genitalia as in Fig. 556.

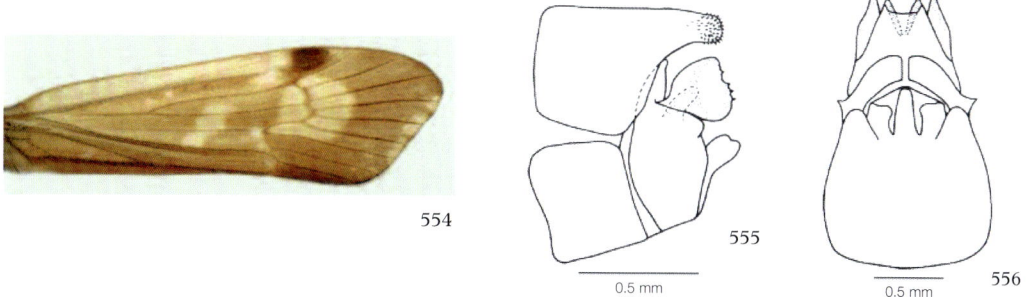

Figures 554-556. *Limnephilus stigma*. 554 fore wing pattern; 555 male genitalia lateral, aedeagus omitted; 556 female genitalia ventral

## *Limnephilus politus* McLachlan, 1865

Fore wing length: ♂♀ 12-17 mm (Figs 557, 558); small irregular markings over much of the wing, but these are absent towards the costal margin. Local in England, Wales and southern Scotland, more widespread in SE. England; no records from Ireland; ponds, lakes and canals. Mainly central and northern Europe. Flight period: August-October. ♂ genitalia as in Fig. 559; ♀ genitalia as in Fig. 560.

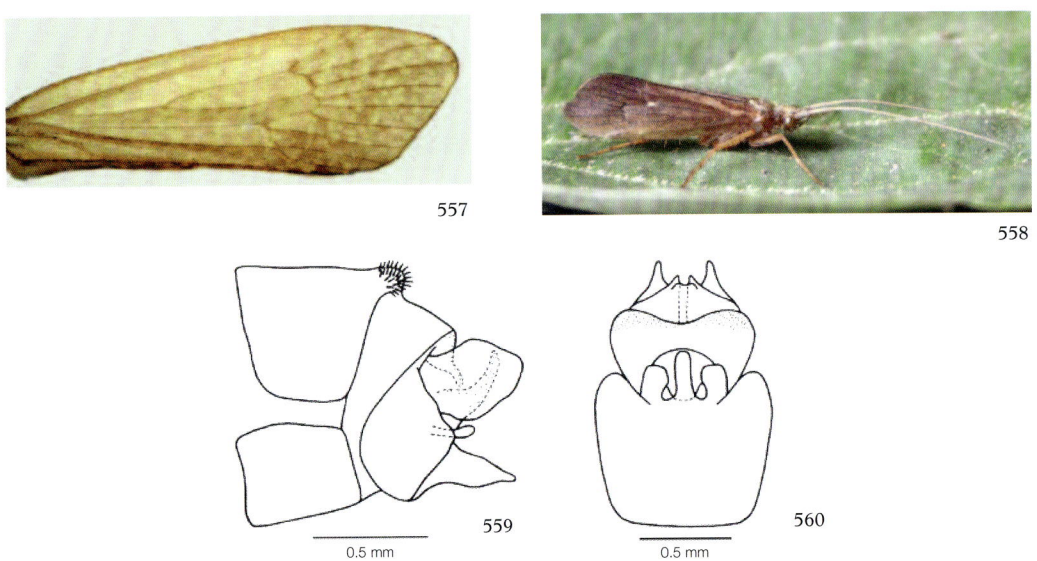

Figures 557-560. *Limnephilus politus.* 557 fore wing pattern; 558 live specimen; 559 male genitalia lateral, aedeagus omitted; 560 female genitalia ventral

## *Limnephilus binotatus* Curtis, 1834
= *xanthodes* McLachlan, 1875

Listed as *L. xanthodes* in Mosely (1939). Fore wing length: ♂♀ 11-16 mm (Fig. 561); the yellow fore wing has variable markings but the pterostigma is always well marked. Local in England and Scotland as far north as Aviemore, only widespread in fens and broads of E. England; present in Ireland; ponds, fens and reed swamps. Throughout Europe except extreme south. Flight period: May-July, with a short diapause. ♂ genitalia as in Fig. 562; ♀ genitalia as in Fig. 563.

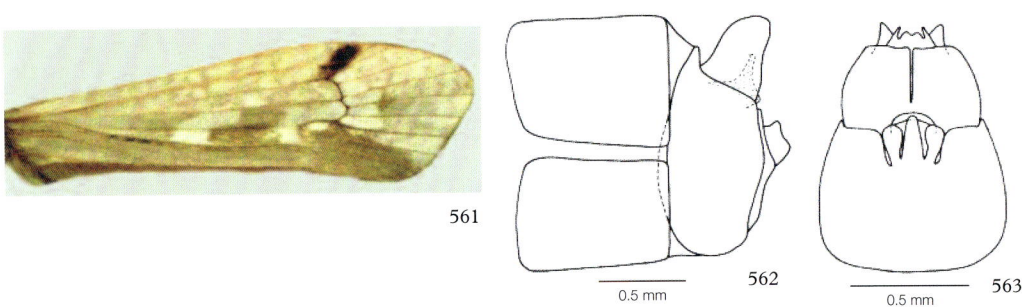

Figures 561-563. *Limnephilus binotatus.* 561 fore wing pattern; 562 male genitalia lateral, aedeagus omitted; 563 female genitalia ventral

## *Limnephilus decipiens* (Kolenati, 1848)

Fore wing length: ♂♀ 12-15 mm (Figs 564, 565); can appear similar to *L. binotatus* but easily distinguished by the genitalia in both sexes. England as far north as south Yorkshire, and Wales, but only widespread in the south and Midlands; present in Ireland; lakes, slow rivers, canals. Throughout most of Europe. Flight period: May-October; as in *L. lunatus* those emerging in the spring probably have a summer diapause, but most emerge in the autumn. ♂ genitalia as in Fig. 566; ♀ genitalia as in Fig. 567.

Figures 564-567. *Limnephilus decipiens*. 564 fore wing pattern; 565 live specimen; 566 male genitalia lateral, aedeagus omitted; 567 female genitalia ventral

## *Limnephilus bipunctatus* Curtis, 1834

Fore wing length: ♂ 12-15 mm, ♀ 12-16 mm (Fig. 568); darkly mottled fore wing but with no distinct pattern, though it can sometimes resemble *L. griseus*. Throughout Britain, but not common; no records from Ireland; habitat unpredictable, but sometimes in small, newly created water bodies. Throughout Europe, less common in north. Flight period: June-October, with summer diapause. ♂ genitalia as in Fig. 569; ♀ genitalia as in Fig. 570.

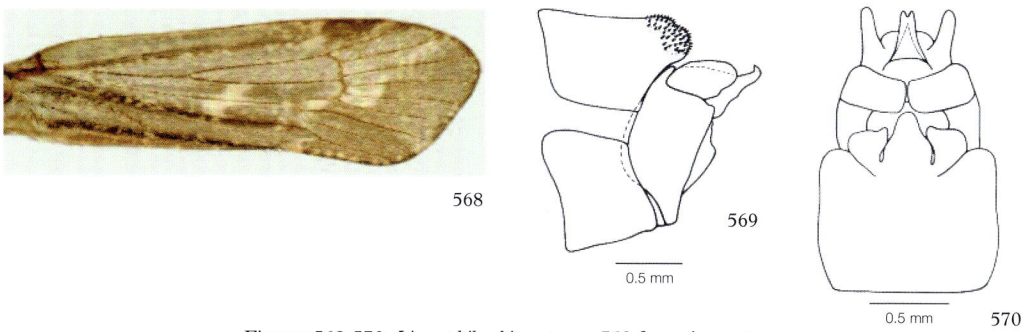

Figures 568-570. *Limnephilus bipunctatus*. 568 fore wing pattern; 569 male genitalia lateral, aedeagus omitted; 570 female genitalia ventral

## *Limnephilus elegans* **Curtis, 1834**

Fore wing length: ♂ 12-15 mm, ♀ 12-16 mm (Fig. 571); very distinctive wing pattern with long dark and pale streaks. Local in New Forest, north Midlands, N. England, Wales and Scotland; present in Ireland; raised bogs and fens. Mainly central and northern Europe. Flight period: May-July. ♂ genitalia as in Fig. 572, the excision in the superior appendages being clearer in dorsal view (Fig. 573); ♀ genitalia as in Fig. 574.

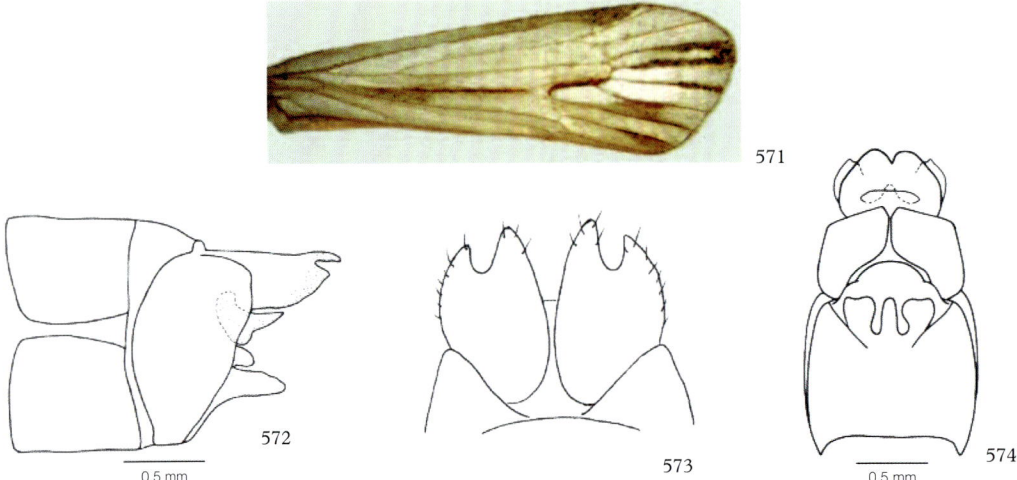

Figures 571-574. *Limnephilus elegans*. 571 fore wing pattern; 572 male genitalia lateral, aedeagus omitted; 573 male genitalia dorsal; 574 female genitalia ventral

## *Limnephilus griseus* **(Linnaeus, 1758)**

Fore wing length: ♂♀ 9-13 mm (Fig. 575); usually very strongly marked. Common throughout Britain, especially upland areas; present in Ireland; pools (often temporary) and marshes, especially acid waters. Throughout Europe. Flight period: April-May, August-October, with summer diapause. ♂ genitalia as in Fig. 576; ♀ genitalia as in Fig. 577.

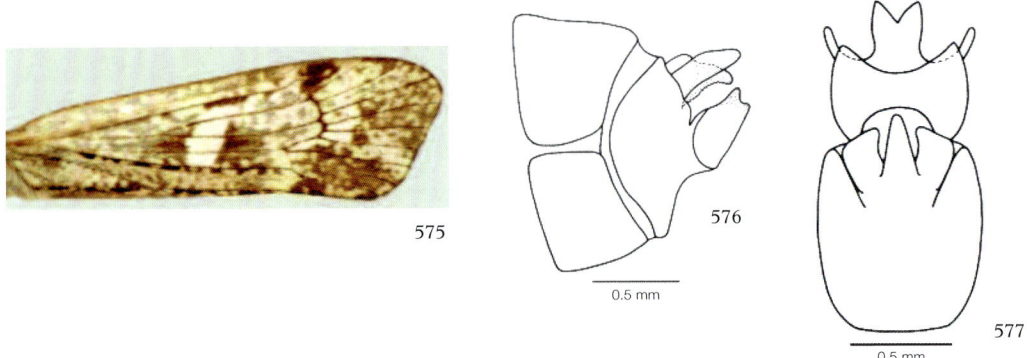

Figures 575-577. *Limnephilus griseus*. 575 fore wing pattern; 576 male genitalia lateral, aedeagus omitted; 577 female genitalia ventral

## *Limnephilus luridus* Curtis, 1834

Fore wing length: ♂ 10-12 mm, ♀ 10-13 mm (Fig. 578); fore wing pale yellowish brown with only faint markings. Common throughout Britain; present in Ireland; temporary pools in woodland or bogs. Mainly northern Europe. Flight period: May-November, with short summer diapause. ♂ genitalia as in Fig. 579; ♀ genitalia as in Fig. 580.

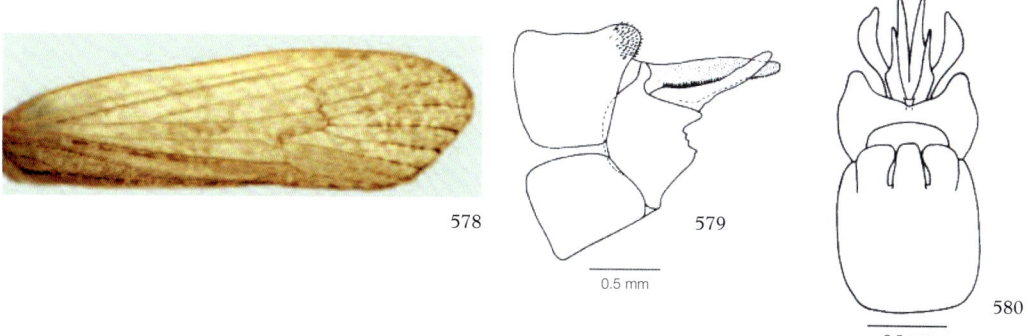

Figures 578-580. *Limnephilus luridus*. 578 fore wing pattern;
579 male genitalia lateral, aedeagus omitted; 580 female genitalia ventral

## *Limnephilus ignavus* McLachlan, 1865

Fore wing length: ♂ 10-12 mm, ♀ 10-15 mm (Fig. 581); fore wing uniformly brown with a small but distinct pale spot at the arculus. Throughout much of Britain, but only common in the north, with few records for S. England and none for Wales; present in Ireland; shallow pools and flowing marshes, fens. Throughout Europe. Flight period: May-June, August-October, with summer diapause. ♂ genitalia as in Fig. 582; ♀ genitalia as in Fig. 583.

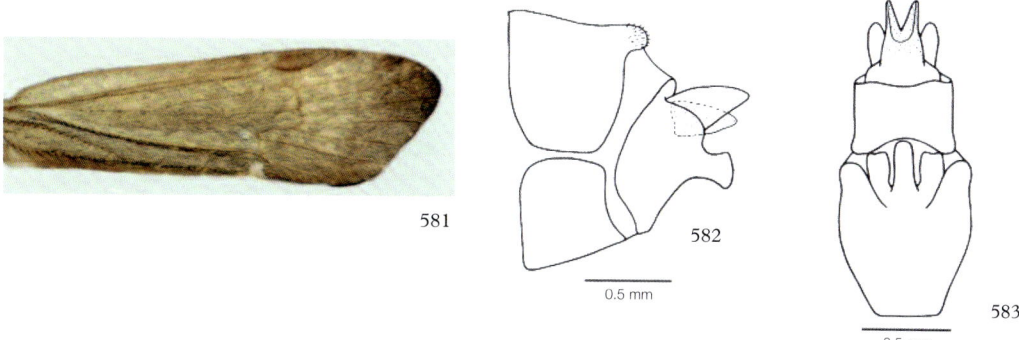

Figures 581-583. *Limnephilus ignavus*. 581 fore wing pattern;
582 male genitalia lateral, aedeagus omitted; 583 female genitalia ventral

## *Limnephilus fuscinervis* (Zetterstedt, 1840)

Fore wing length: ♂ 10-12 mm, ♀ 10-14 mm (Fig. 584); pale yellow but with conspicuous veins. Ireland only. Mainly northern Europe. ♂ genitalia as in Fig. 585; ♀ genitalia as in Fig. 586.

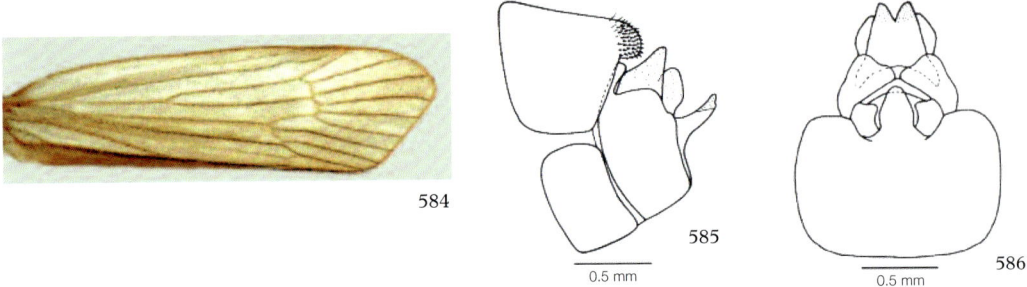

Figures 584-586. *Limnephilus fuscinervis*. 584 fore wing pattern;
585 male genitalia lateral, aedeagus omitted; 586 female genitalia ventral

## *Limnephilus sparsus* Curtis, 1834

Fore wing length: ♂ 10-12 mm, ♀ 10-13 mm (Fig. 587); very dark brown with white spots near the pterostigma and arculus. Common throughout Britain; present in Ireland; temporary puddles, pools and marshes. Throughout Europe. Flight period: May-June, August-October, with summer diapause. ♂ genitalia as in Fig. 588; ♀ genitalia as in Fig. 589.

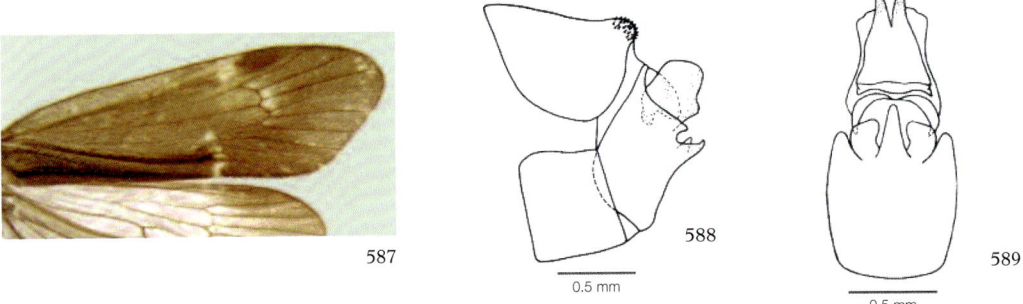

Figures 587-589. *Limnephilus sparsus*. 587 fore wing pattern;
588 male genitalia lateral, aedeagus omitted; 589 female genitalia ventral

## *Limnephilus fuscicornis* (Rambur, 1842)

Fore wing length: ♂ 14-16 mm, ♀ 14-17 mm (Fig. 590) fore wing noticeably broad, uniformly blackish when alive, fading to brown after death. England, Wales and S. Scotland, but not common; no records from Ireland; rivers and streams. Throughout Europe, except south. Flight period: May-August. A retiring species that hides in crevices in tree-bark and is usually not attracted to light. ♂ genitalia as in Fig. 591; ♀ genitalia as in Fig. 592.

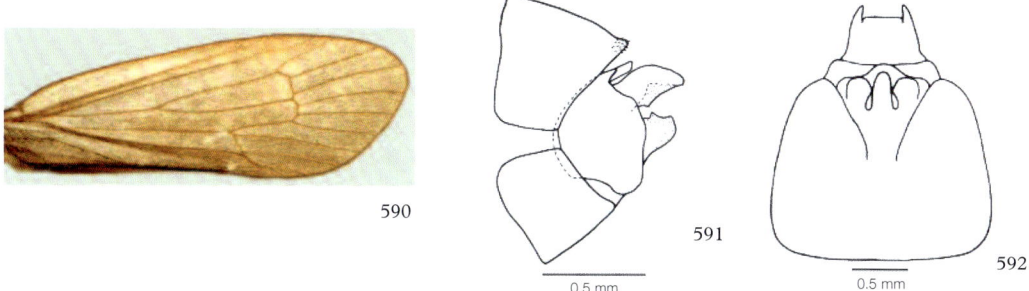

Figures 590-592. *Limnephilus fuscicornis.* 590 fore wing pattern;
591 male genitalia lateral, aedeagus omitted; 592 female genitalia ventral

## *Limnephilus extricatus* McLachlan, 1865

Fore wing length: ♂♀ 12-14 mm (Figs 593); usually uniformly brown with some darker areas, and can resemble *L. hirsutus*. Common throughout much of Britain, though only few records for S. and SW. England; no records from Ireland; slow flowing water. Throughout most of Europe. Flight period: June-September. ♂ genitalia as in Fig. 594; ♀ genitalia as in

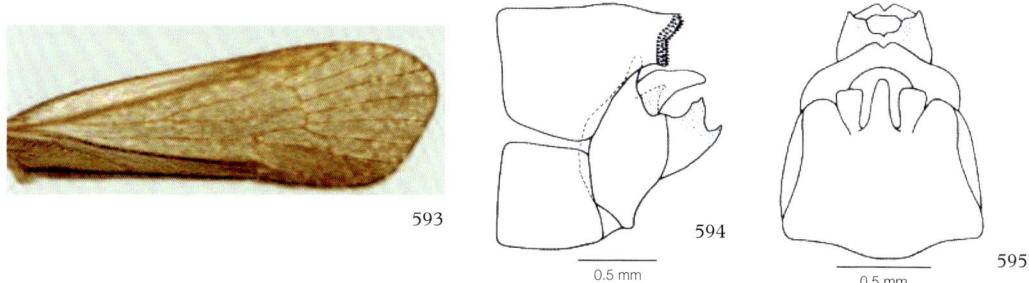

Figures 593-595. *Limnephilus extricatus.* 593 fore wing pattern;
594 male genitalia lateral, aedeagus omitted; 595 female genitalia ventral

Fig. 595.
## *Limnephilus centralis* Curtis, 1834

Fore wing length: ♂♀ 8-10 mm: very variable, with the typical pale central streak (Fig. 596) sometimes replaced by general mottling (Fig. 597). Common throughout Britain; present in Ireland; small upland trickles and small (usually temporary) ditches and pools. Throughout Europe. Flight period: May-September, with summer diapause. ♂ genitalia as in Fig. 598;

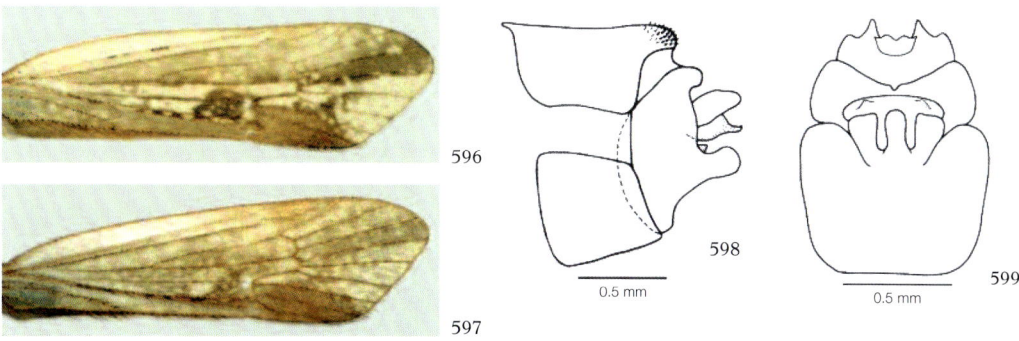

Figures 596-599. *Limnephilus centralis.* 596, 597 variations in fore wing pattern;
598 male genitalia lateral, aedeagus omitted; 599 female genitalia ventral

♀ genitalia as in Fig. 599.
## *Limnephilus hirsutus* (Pictet, 1834)

Fore wing length: ♂♀ 10-12 mm (Figs 600, 601): the very long discoidal cell in the fore wing, usually at least twice the length of its stalk, separates it from *L. extricatus*, in which the cell is not more than 1.25 times the length of the stalk. *L. extricatus*, *L. pati* and *L. tauricus* all have a similar pattern wing pattern. The relative lengths of the discoidal cell in *L. hirsutus*, *L. pati* and *L. tauricus* seem to give a clear separation, but this character should be used with caution because it has been tested on very few specimens. Local but widespread throughout Britain; present in Ireland; small bare muddy trickles and ditches, sometimes in unstable areas such as cliffs and claypits. Throughout much of Europe. Flight period: April-September, possibly with a diapause. ♂ genitalia as in Fig. 602; ♀ genitalia as in Fig. 603.

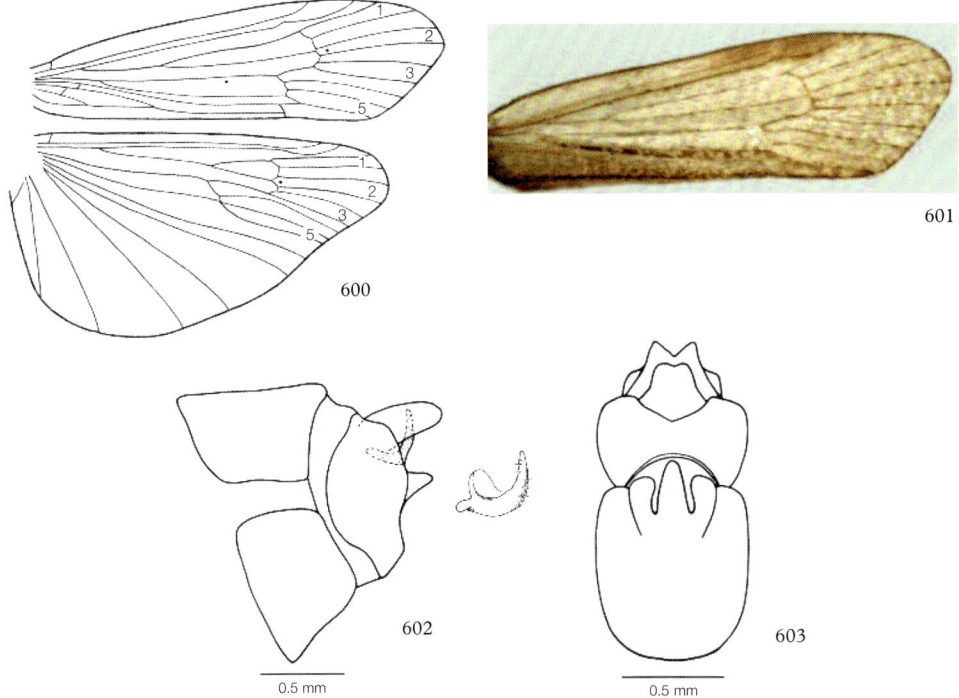

Figures 600-603. *Limnephilus hirsutus*. 600 wing venation; 601 fore wing pattern;
602 male genitalia lateral, aedeagus omitted, intermediate appendage shown separately; 603 female genitalia ventral

## *Limnephilus pati* O'Connor, 1980

This species was not distinguished from *L. hirsutus* until the description by O'Connor (1980). Fore wing length: ♂ ♀ 10-12 mm, with a similar pattern to *L. hirsutus* and *L. tauricus*. The discoidal cell in the fore wing is at least 1.75 times the length of its stalk, though not as long as in *L. hirsutus*. Early records from E. England, but recent records only from the Isle of Man; present in Ireland; fens. Outside the British Isles currently only recorded from Germany, but it may prove to be more widespread when all records of the *L. hirsutus* group are reassessed. Flight period: June-July. ♂ genitalia as in Fig. 604; ♀ genitalia as in Fig. 605.

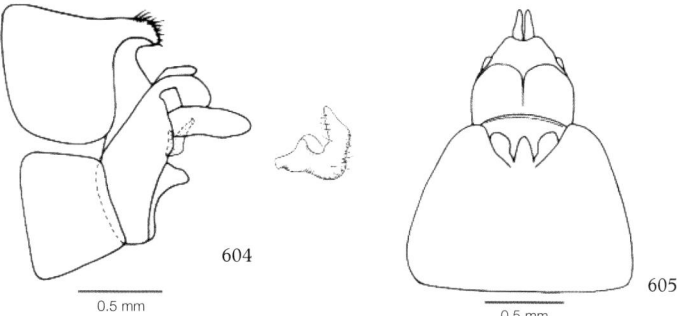

Figures 604-605. *Limnephilus pati.* 604 male genitalia lateral, aedeagus omitted, intermediate appendage shown separately; 605 female genitalia ventral

## *Limnephilus tauricus* Schmid, 1964

Not distinguished as a British species until O'Connor & Barnard (1981). Fore wing length: ♂ 10-13 mm, ♀ 10-14 mm. Fore wing discoidal cell is around 1.6 times the length of its stalk, not as long as in *L. hirsutus* or *L. pati*, with which it shares a similar wing pattern. Early records from E. England, recent records only from Berkshire, Hampshire and Anglesey; present in Ireland; fens. Scattered records throughout central Europe. Flight period: July-August. ♂ genitalia as in Fig. 606; ♀ genitalia as in Fig. 607.

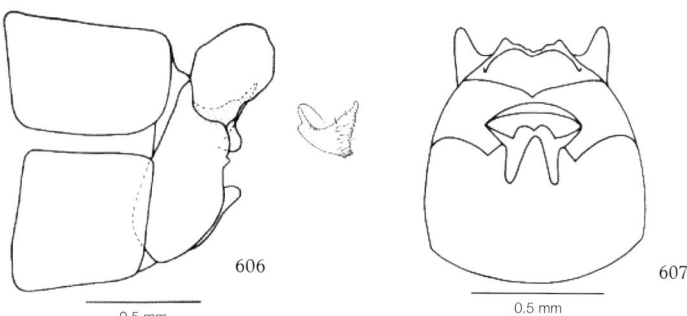

Figures 606-607. *Limnephilus tauricus.* 606 male genitalia lateral, aedeagus omitted, intermediate appendage shown separately; 607 female genitalia ventral

## *Limnephilus auricula* **Curtis, 1834**

Fore wing length: ♂ ♀ 8-12 mm (Figs 608, 609); fore wing a very distinctive golden brown with pale markings, hind wing with darkened area at apex. Common throughout Britain; present in Ireland; temporary pools, ponds and ditches. Throughout Europe. Flight period: (April) May-October (November), with summer diapause. ♂ genitalia as in Fig. 610; ♀ genitalia as in Fig. 611.

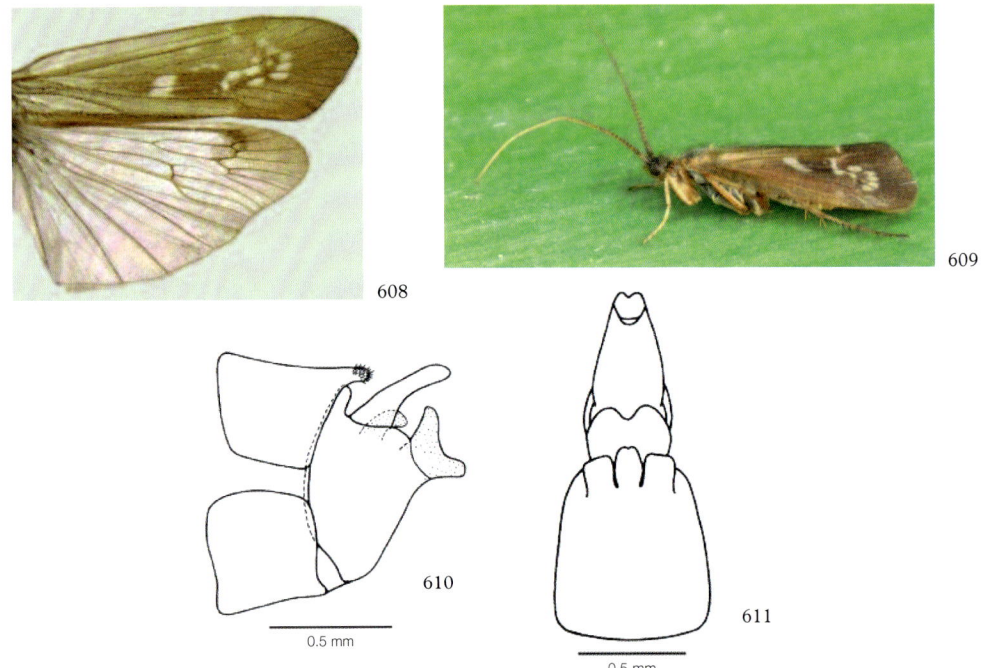

Figures 608-611. *Limnephilus auricula*. 608 wing pattern; 609 live specimen;
610 male genitalia lateral, aedeagus omitted; 611 female genitalia ventral

## *Limnephilus nigriceps* **(Zetterstedt, 1840)**

Fore wing length: ♂ 10-13 mm, ♀ 10-14 mm (Fig. 612); fore wing with only indistinct markings. Scattered records from England, more widespread in Scotland north to Speyside; present in Ireland; lakes and large ponds. Mainly central and northern Europe. Flight period: September-October; the late flight period may account for the scarcity of records. ♂ genitalia as in Fig. 613; ♀ genitalia as in Fig. 614.

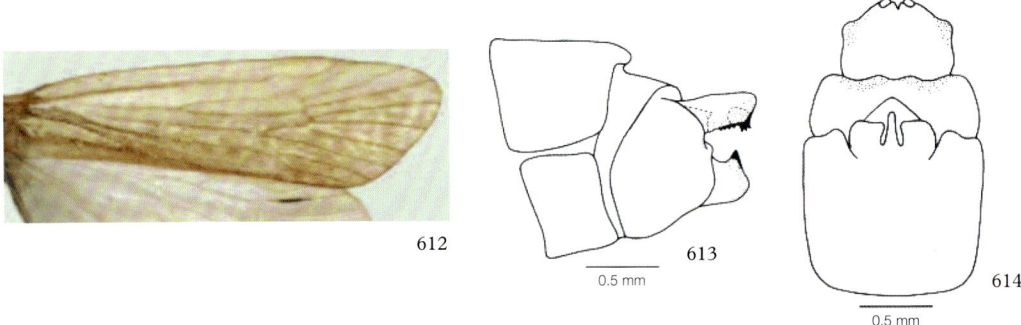

Figures 612-614. *Limnephilus nigriceps*. 612 fore wing pattern;
613 male genitalia lateral, aedeagus omitted; 614 female genitalia ventral

## *Limnephilus coenosus* **Curtis, 1834**

Listed as *Asynarchus coenosus* in Mosely (1939). Fore wing length: ♂♀ 10-13 mm (Figs 615); fore wing broad, brown with a few distinct white spots. Scattered records from SW. England, Wales and the bordering English counties, more common in N. England and Scotland; present in Ireland; permanent small pools on moorland and bogs. Throughout much of Europe. Flight period: July-October. ♂ genitalia as in Fig. 616; ♀ genitalia as in Fig. 617.

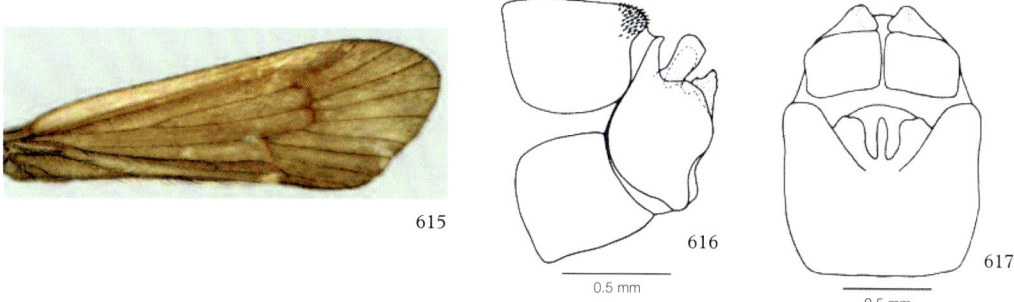

Figures 615-617. *Limnephilus coenosus.* 615 fore wing pattern;
616 male genitalia lateral, aedeagus omitted; 617 female genitalia ventral

## *Limnephilus vittatus* **(Fabricius, 1798)**

Fore wing length: ♂♀ 8-11 mm (Figs 618-620); fore wing with distinctive pattern, predominantly pale in anterior half and dark in posterior half; distinguished from all other species in the genus by fork 3 in fore wing and hind wing usually being stalked. Common throughout Britain; present in Ireland; bare-bottomed pools, ditches (often temporary) and reservoirs. Throughout Europe. Flight period: May-October, with summer diapause. ♂ genitalia as in Fig. 621; ♀ genitalia as in Fig. 622.

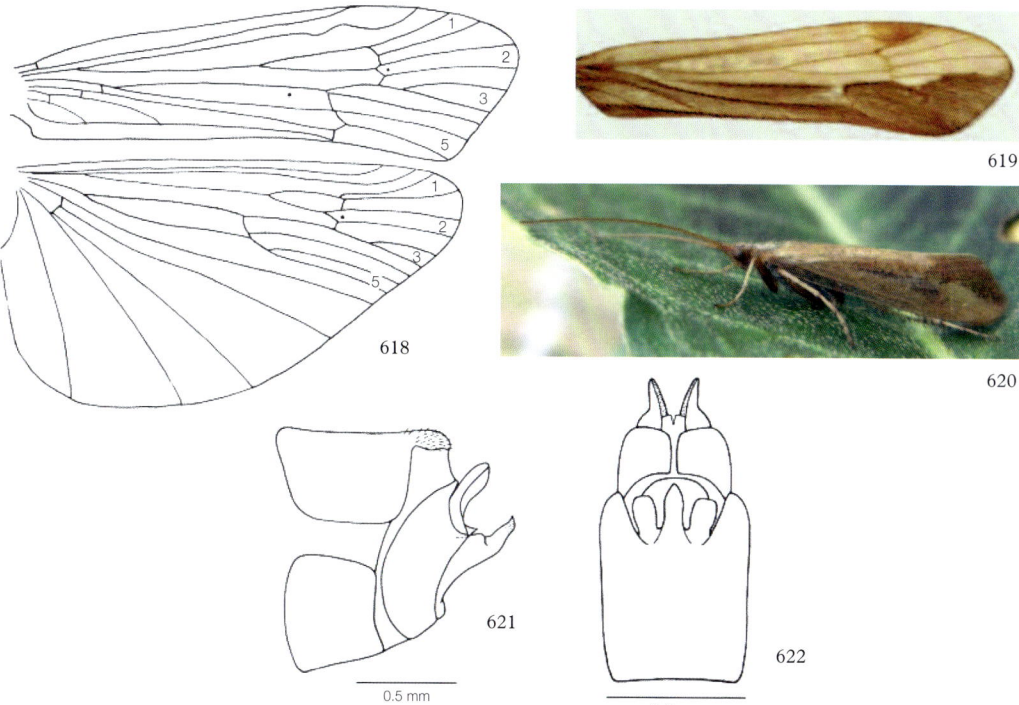

Figures 618-622. *Limnephilus vittatus.* 618 wing venation; 619 fore wing pattern; 620 live specimen;
621 male genitalia lateral, aedeagus omitted; 622 female genitalia ventral

# Tribe Chaetopterygini

## Genus CHAETOPTERYX Stephens, 1829

Spur formula 0.3.3 in male, 1.3.3 in female. Of around 25 European species only one occurs in Britain.

### *Chaetopteryx villosa* (Fabricius, 1798)

Fore wing length: ♂ 6-11 mm, ♀ 7-12 mm (Figs 623-626); fore wing very broad and granular with long bristles (only *Anabolia brevipennis* has a similar appearance, but its wings are not as broad and rounded, compare Fig. 497). Common throughout Britain; present in Ireland; streams, rivers, small upland lakes. Throughout most of western and northern Europe. Flight period: (August) September-November. ♂ genitalia as in Fig. 627; ♀ genitalia as in Fig. 628.

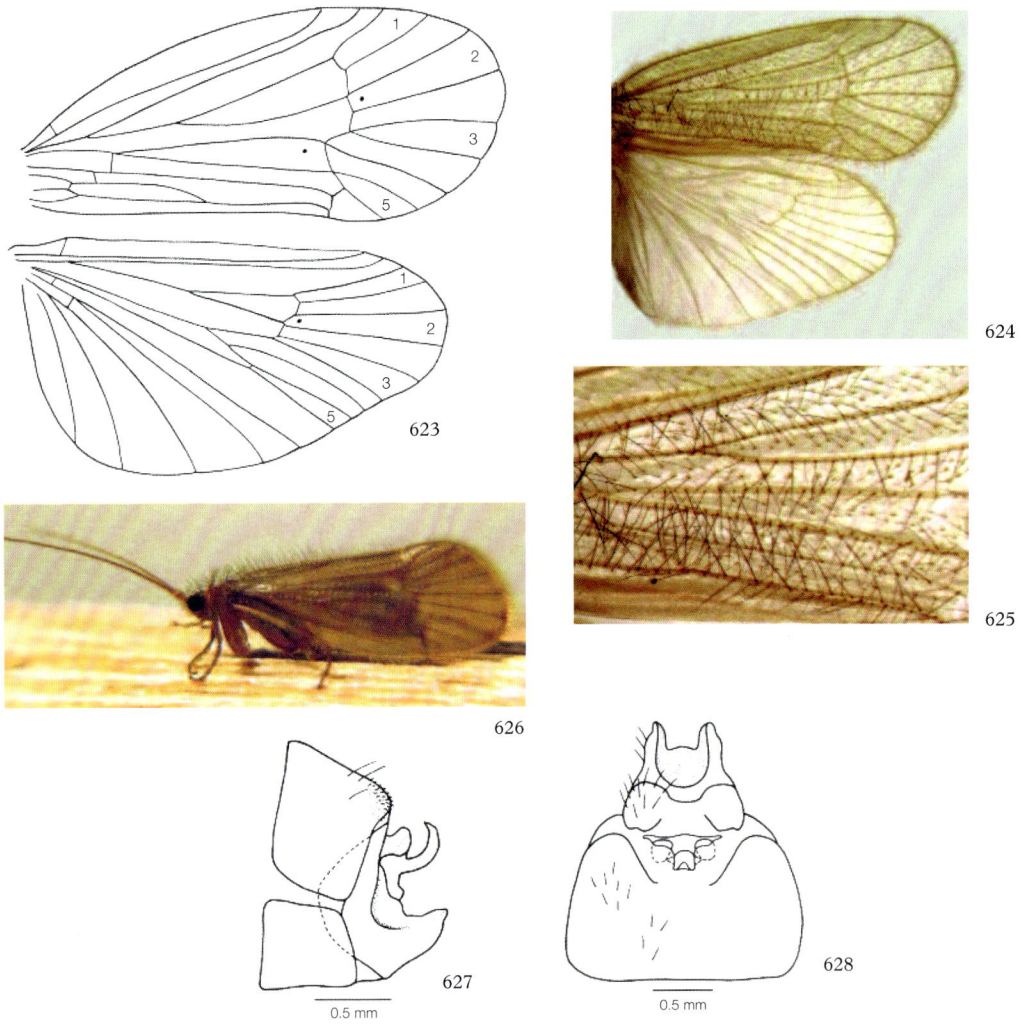

Figures 623-628. *Chaetopteryx villosa*. 623 wing venation; 624 wing pattern; 625 closeup of fore wing;
626 live specimen [photo: Peter Barnard]; 627 male genitalia lateral, aedeagus omitted; 628 female genitalia ventral

# Tribe Stenophylacini

## Genus MESOPHYLAX McLachlan, 1882

Spur formula 1.3.4 (spur on fore tibia of male very small). Two of the six European species are found in Britain.

### *Mesophylax aspersus* (Rambur, 1842)

Fore wing length: ♂ 15-16 mm, ♀ 15-20 mm (Fig. 629): the wings are usually darker than in *M. impunctatus*, though the genitalia give a more reliable separation. Only three records from E. and S. England, presumed migrants; no records from Ireland; would probably breed in small temporary streams. Mainly south-western Europe. Flight period: February-April, October. ♂ genitalia with small projection on posterior surface of clasper (Fig. 630); ♀ genitalia with shallow rounded apical incision on central vulvar scale (Fig. 631).

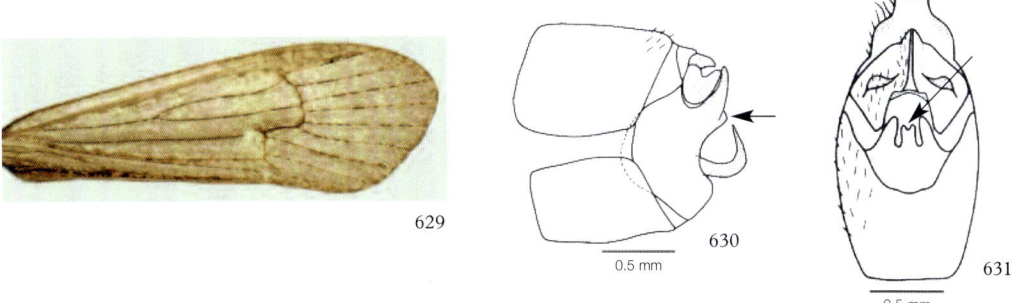

Figures 629-631. *Mesophylax aspersus*. 629 fore wing pattern;
630 male genitalia lateral, aedeagus omitted; 631 female genitalia ventral

### *Mesophylax impunctatus* McLachlan, 1884

All British specimens belong to the subspecies *M. impunctatus zetlandicus*. There are two other subspecies in Europe (Malicky, 1998; 2004). Fore wing length: ♂ 12-17 mm, ♀ 14-18 mm (Fig. 632); fore wing speckled as in *M. aspersus*. Malham Tarn, NW. England and Scotland; present in Ireland; lakes. Flight period: May-November, with summer diapause. ♂ genitalia with no projection on posterior surface of clasper (Fig. 633); ♀ genitalia with deep V-shaped apical incision on central vulvar scale (Fig. 634).

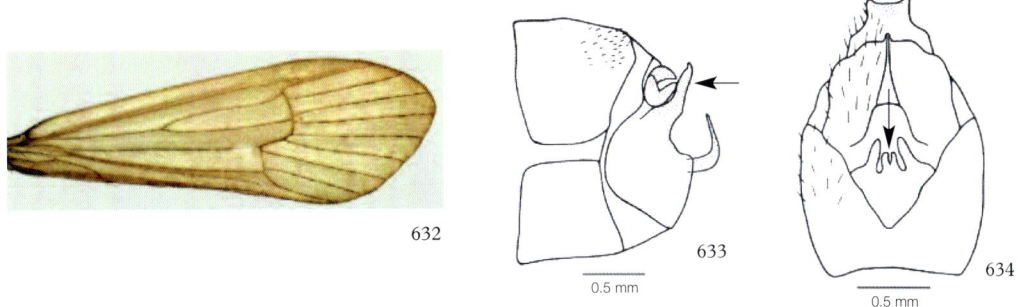

Figures 632-634. *Mesophylax impunctatus*. 632 fore wing pattern;
633 male genitalia lateral, aedeagus omitted; 634 female genitalia ventral

# Genus MICROPTERNA Stein, 1874

This genus is closely related to *Stenophylax*, and the differences in both adult and larval characters are small. Around 20 species in Europe, with just two in Britain. Spur formula 0.3.4 in male, 1.3.4 in females.

## *Micropterna lateralis* (**Stephens, 1837**)

Listed as *Stenophylax lateralis* in Macan (1973). Fore wing length: ♂ ♀ 14-18 mm (Fig. 635); rather uniformly brown fore wing with some speckling. Common throughout Britain; present in Ireland; ditches and temporary streams. Throughout much of Europe except extreme south. Flight period: (May) June-August. ♂ genitalia with broad rounded superior appendages, and clasper ending in a sharp upwardly directed point (Fig. 636); ♀ genitalia with prominent apex to segment X, visible between terminal lobes in ventral view (Fig. 637).

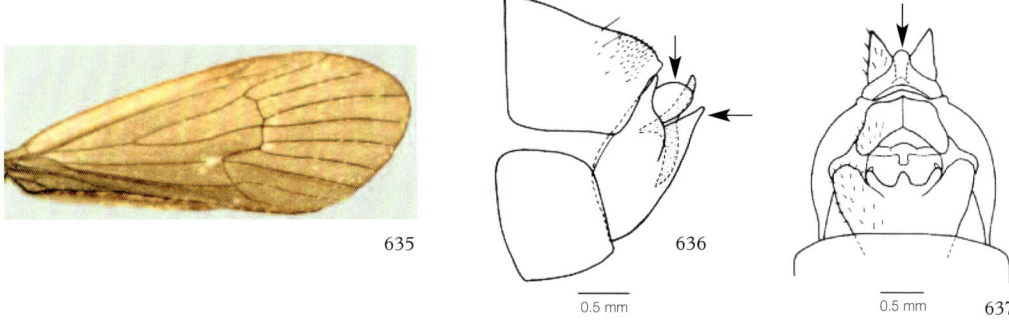

Figures 635-637. *Micropterna lateralis.* 635 fore wing pattern;
636 male genitalia lateral, aedeagus omitted; 637 female genitalia ventral

## *Micropterna sequax* **McLachlan, 1875**

Listed as *Stenophylax sequax* in Macan (1973). Fore wing length: ♂ 13-18 mm, ♀ 14-18 mm (Figs 638-640); wing pattern similar to *M. lateralis* but with pale areas in the apical forks. Common throughout Britain; present in Ireland; temporary or semi-permanent streams. Throughout Europe. Flight period: (May) June-October, possibly with summer diapause. ♂ genitalia with small and narrow superior appendages, clasper with no terminal pointed section (Fig. 641); ♀ genitalia with short segment X, not visible in ventral view (Fig. 642).

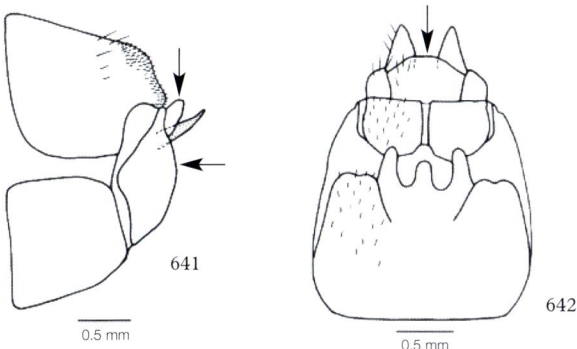

Figures 638-642. *Micropterna sequax.* 638 wing venation; 639 fore wing pattern; 640 live specimen; 641 male genitalia lateral, aedeagus omitted; 642 female genitalia ventral

# Genus STENOPHYLAX Kolenati, 1848

Spur formula 1.3.4. Of around 10 European species two occur in Britain. Together with the genus *Potamophylax* these are the angler's Large Cinnamon Sedges.

### *Stenophylax permistus* McLachlan, 1895
= *concentricus* McLachlan, 1875

Fore wing length: ♂ 18-22 mm, ♀ 20-24 mm (Fig. 643); yellowish brown with darker speckles all over. Common throughout Britain; present in Ireland; temporary water including ditches, woodland streams and marshes. Throughout Europe. Flight period: (March) April-June, August-November, with summer diapause. This species disperses widely from its breeding sites and is found in light-traps almost anywhere. ♂ genitalia with claspers hardly separated from segment IX, even at apex (Fig. 644); ♀ genitalia with segment X small and bifurcated apically, bearing two small pointed processes directed posteriorly (Fig. 645).

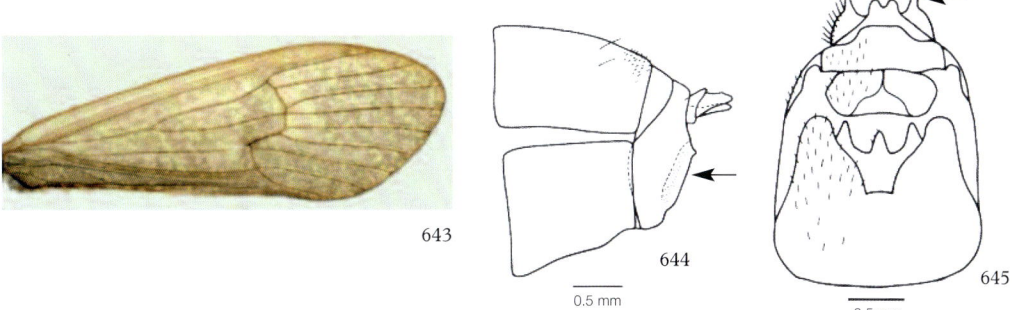

Figures 643-645. *Stenophylax permistus.* 643 fore wing pattern; 644 male genitalia lateral, aedeagus omitted; 645 female genitalia ventral

### *Stenophylax vibex* (Curtis, 1834)

Fore wing length: ♂♀ 18-23 mm (Figs 646, 647); similar to *S. permistus* but with no speckling anterior to the radial vein. Local throughout much of Britain, though absent from many counties; no records from Ireland; small temporary streams. Mainly southern and central Europe. Flight period: May-June, August-November, with summer diapause. ♂ genitalia with apex of claspers clearly separate from segment IX (Fig. 648); ♀ genitalia with segment X larger than in *S. permistus* and with the terminal processes larger and divergent (Fig. 649).

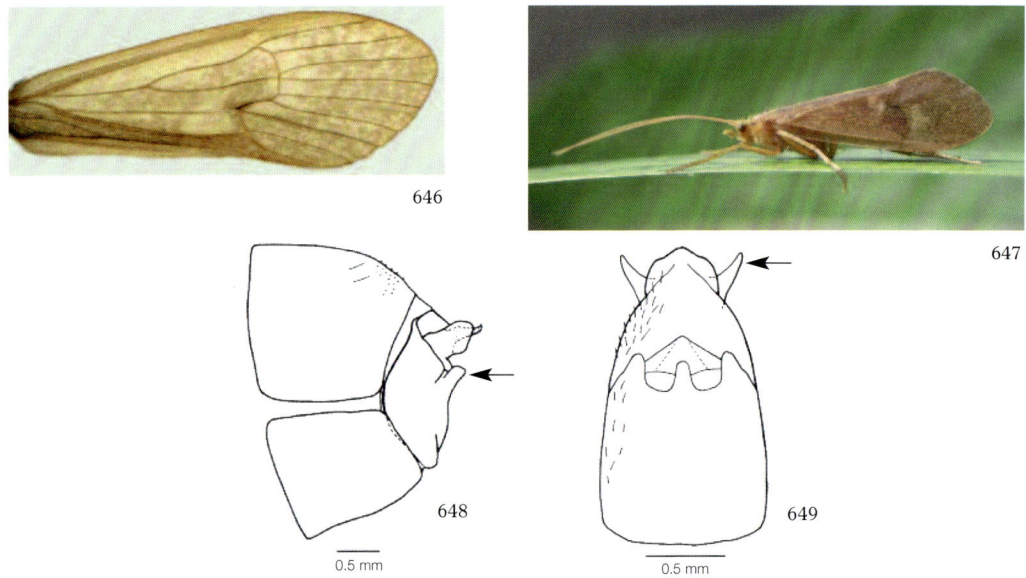

Figures 646-649. *Stenophylax vibex*. 646 fore wing pattern;
647 live specimen; 648 male genitalia lateral, aedeagus omitted; 649 female genitalia ventral

## Genus POTAMOPHYLAX Wallengren, 1891

Spur formula 1.3.4. Of around 15 European species three occur in Britain. With the genus *Stenophylax* these are the angler's Large Cinnamon Sedges. In dead and dried specimens the superior appendages of the male genitalia often contract and are difficult to see. After clearing, or at least softening the genitalia, these appendages must be viewed flat on, not obliquely, so that the true outline can be compared with the figures.

### *Potamophylax cingulatus* (Stephens, 1837)
= *latipennis*; misidentified by some authors

Listed as *Stenophylax latipennis* in Mosely (1939). Fore wing length: ♂♀ 14-20 mm (Fig. 650); rather greyish yellow with paler markings. Common throughout Britain, though local in S. England; present in Ireland; streams and rivers, sometimes lakes in N. Scotland. Throughout most of Europe. Flight period: (June) July-October. ♂ genitalia with the two lobes of the superior appendages of very different sizes, the inner lobe being much smaller (Fig. 651); ♀ genitalia as in Fig. 652. The females cannot be separated reliably from those of *P. latipennis*, despite attempts such as Décamps (1966) and Robert & Schmidt (1990). Any apparent differences in female genitalia do not seem to be consistent.

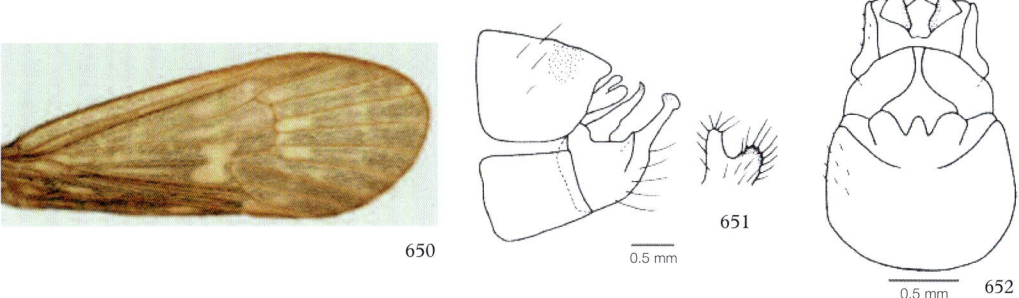

Figures 650-652. *Potamophylax cingulatus.* 650 fore wing pattern;
651 male genitalia lateral, aedeagus omitted; tip of left superior appendage, dorsal view;
652 Potamophylax sp. female genitalia ventral

## *Potamophylax latipennis* (Curtis, 1834)
= *stellatus* (Curtis, 1834)

Listed as *Stenophylax stellatus* in Mosely (1939). Fore wing length: ♂ ♀ 14-20 mm (Figs 653-655); coloration similar to *P. cingulatus.* Common throughout Britain; present in Ireland; streams, rivers and lakes. Throughout most of Europe. Flight period: June-October. ♂ genitalia with the two lobes of the superior appendages both rounded and of a similar size (Fig. 656); ♀ genitalia as in Fig. 652. The females cannot be reliably separated from those of *P. cingulatus.*

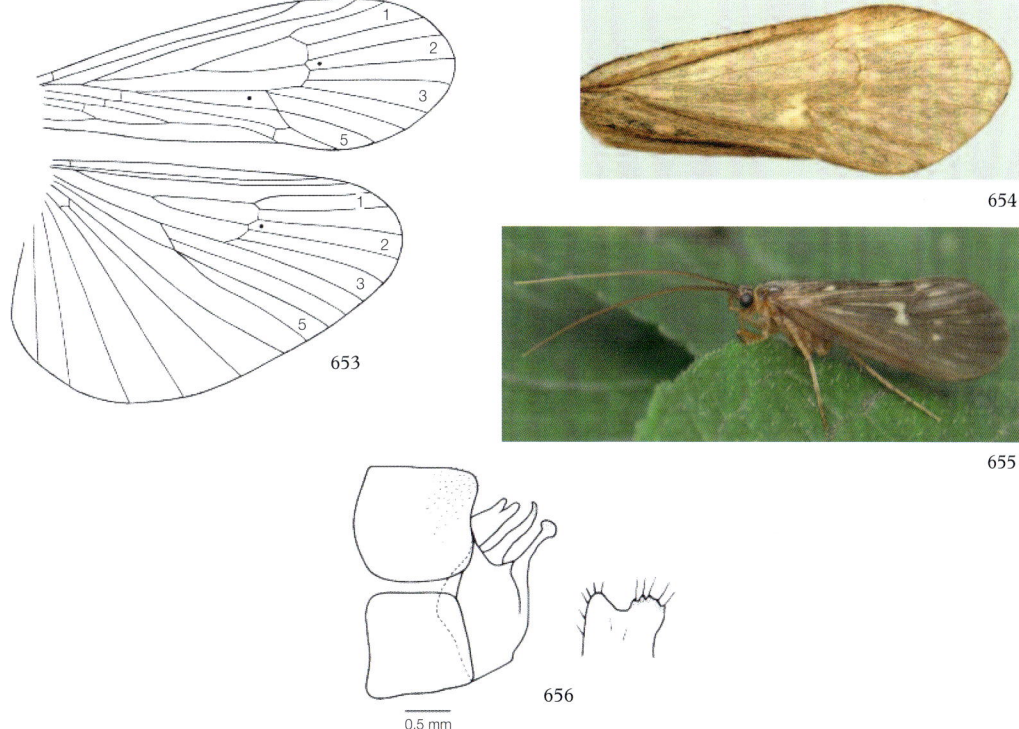

Figures 653-656. *Potamophylax latipennis.* 653 wing venation; 654 fore wing pattern; 655 live specimen;
656 male genitalia lateral, aedeagus omitted; tip of left superior appendage, dorsal view

### *Potamophylax rotundipennis* (Brauer, 1857)

Listed as *Stenophylax rotundipennis* in Mosely (1939). Fore wing length: ♂♀ 14-17 mm (Fig. 657); coloration similar to the two other species of *Potamophylax*, but fore wing rather broader. Throughout Britain, but very local and absent from many counties; no records from Ireland; streams and small rivers. Throughout much of Europe except south. Flight period: August-September. ♂ genitalia as in Fig. 658; ♀ genitalia with three narrow pointed processes on segment X (Fig. 659).

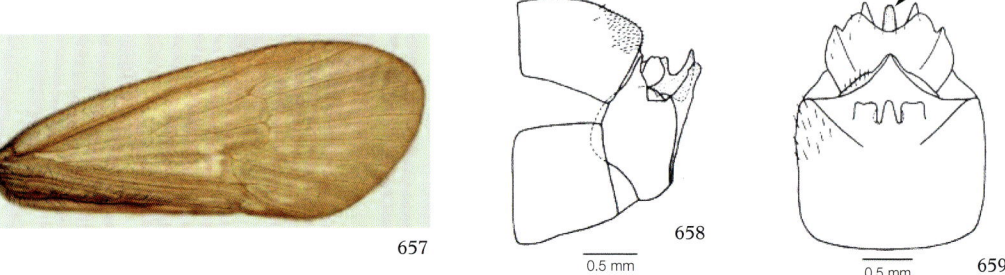

Figures 657-659. *Potamophylax rotundipennis*. 657 fore wing pattern; 658 male genitalia lateral, aedeagus omitted; 659 female genitalia ventral

# Genus ALLOGAMUS Schmid, 1955

Spur formula 1.3.3. Of about 15 European species just one occurs in Britain.

### *Allogamus auricollis* (Pictet, 1834)

Listed as *Halesus auricollis* in Mosely (1939). Fore wing length: ♂♀ 12-16 mm (Figs 660, 661); dark brown with darker pterostigma and pale spot at thyridium. Local in Herefordshire and Gloucestershire, Staffordshire northwards, and mainland Scotland; no records from Ireland; stony rivers and large streams. Mainly southern and central Europe. Flight period: July-October. ♂ genitalia as in Fig. 662; ♀ genitalia as in Fig. 663.

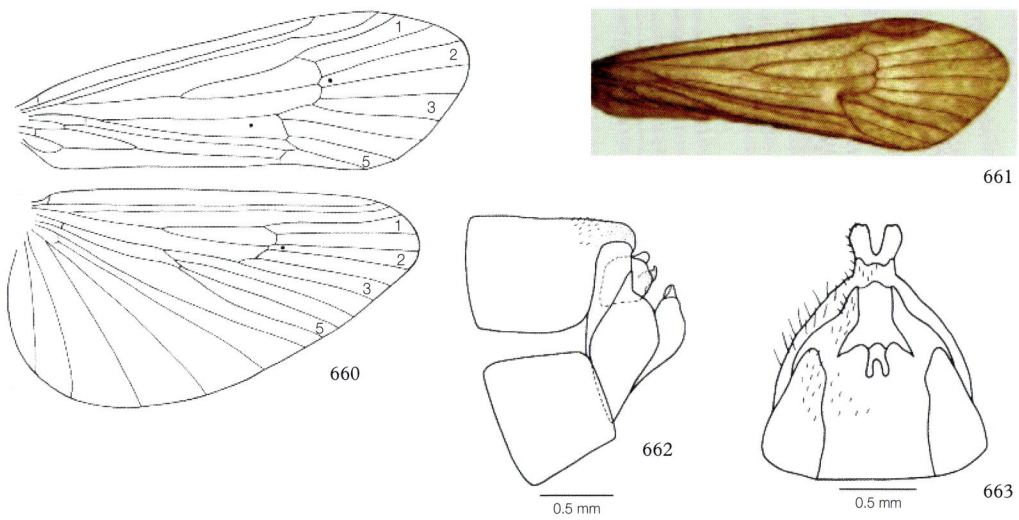

Figures 660-663. *Allogamus auricollis*. 660 wing venation; 661 fore wing pattern; 662 male genitalia lateral, aedeagus omitted; 663 female genitalia ventral

# Genus HALESUS Stephens, 1836

Spur formula 1.3.3. Eight species occur in Europe with two in Britain. These are the angler's Caperers. As in *Potamophylax*, in dead and dried specimens the superior appendages of the male genitalia often contract and are difficult to see.

## *Halesus digitatus* (Schrank, 1781)

Fore wing length: ♂♀ 17-23 mm (Figs 664-665); the dark speckling in the apical forks rather irregularly branched. Common throughout England, Wales and mainland Scotland; present in Ireland; small streams and rivers. Throughout most of Europe. Flight period: (August) September-November. ♂ genitalia as in Fig. 666; ♀ genitalia with apex of segment X broadly triangular (Fig. 667).

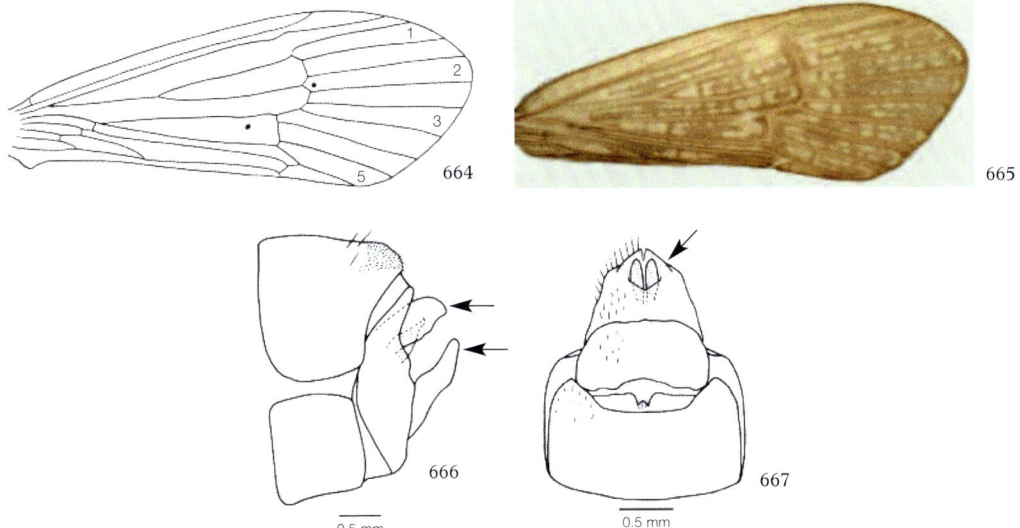

Figures 664-667. *Halesus digitatus*. 664 wing venation; 665 fore wing pattern; 666 male genitalia lateral, aedeagus omitted; 667 female genitalia ventral

## *Halesus radiatus* (Curtis, 1834)

Fore wing length: ♂♀ 17-23 mm (Figs 668, 669); the dark speckling in the apical forks forming well-defined long streaks. Common throughout Britain; present in Ireland; streams, rivers, lakes. Throughout most of Europe. Flight period: (July) August-November. ♂ genitalia as in Fig. 670; ♀ genitalia with segment X projecting, more parallel-sided than in *H. digitatus* and with deeper apical incision (Fig. 671).

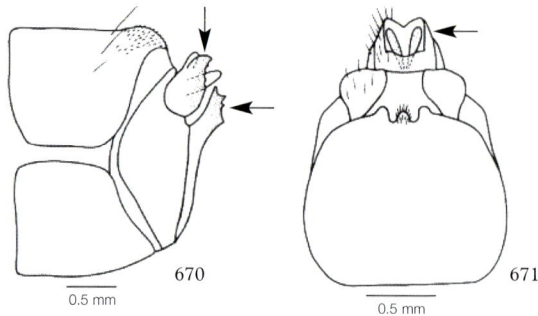

Figures 668-671. *Halesus radiatus*. 668 fore wing pattern; 669 live specimen;
670 male genitalia lateral, aedeagus omitted; 671 female genitalia ventral

## Genus MELAMPOPHYLAX Schmid, 1955

Spur formula 1.3.3. Of around eight species in Europe just one occurs in Britain.

### *Melampophylax mucoreus* (Hagen, 1861)
= *guttatipennis* McLachlan, 1865

Listed as *Halesus guttatipennis* in Mosely (1939). Fore wing length: ♂♀ 10-15 mm (Fig. 672); fairly uniform dark speckling. Very local in England, and a few records from Wales and Scotland; no records from Ireland; alkaline rivers, streams and Malham Tarn. Mainly southern and central Europe. Flight period: (August) September-October (November). ♂ genitalia as in Fig. 673; ♀ genitalia as in Fig. 674.

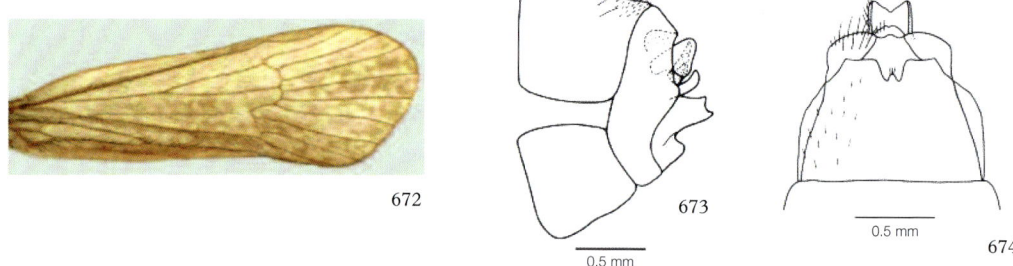

Figures 672-674. *Melampophylax mucoreus*. 672 fore wing pattern;
673 male genitalia lateral, aedeagus omitted; 674 female genitalia ventral

## Genus ENOICYLA Rambur, 1842

Spur formula 0.2.2. The female is the only apterous caddisfly in Britain. One species in Britain out of three found in Europe.

### *Enoicyla pusilla* (Burmeister, 1839)

Fore wing length: ♂ 5-6 mm (Figs 675, 676); greyish brown with no patterning; ♀ apterous (Fig. 677). Known only from the Wyre Forest region (Herefordshire, Worcestershire and Shropshire); no records from Ireland; terrestrial larva in damp oak litter (Green &

Westwood, 2005; Whitehead, 2007). Mainly southern and central Europe. Flight period: June-October, possibly bivoltine. ♂ genitalia as in Fig. 678; ♀ genitalia as in Fig. 679.

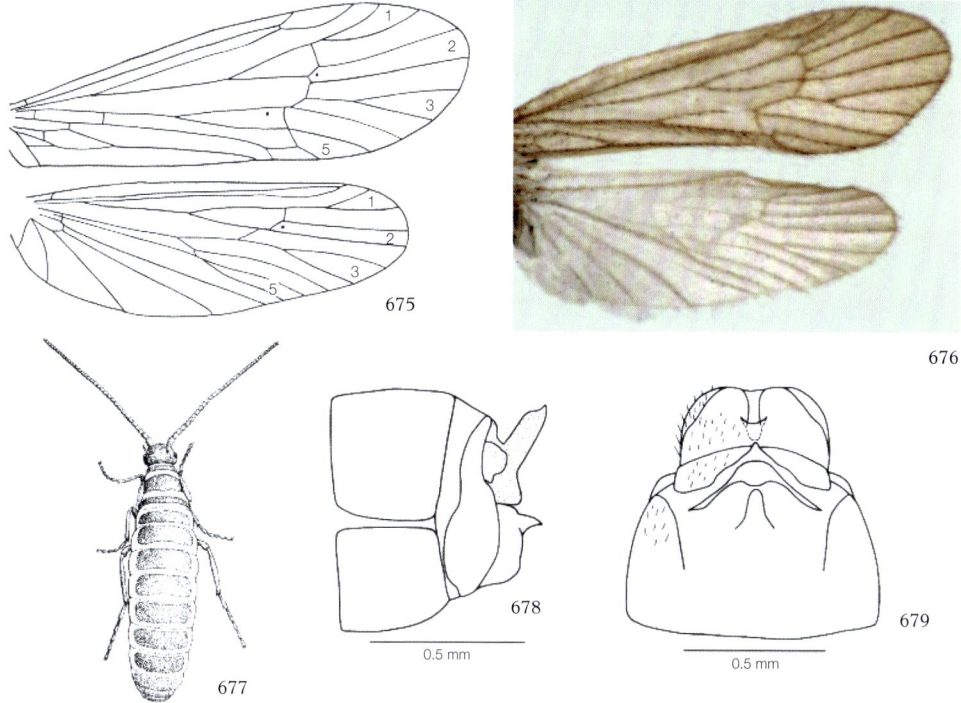

Figures 675-679. *Enoicyla pusilla.* 675 wing venation; 676 wing pattern; 677 female (from Rathjen, 1939); 678 male genitalia lateral, aedeagus omitted; 679 female genitalia ventral

# Genus HYDATOPHYLAX Wallengren, 1891

Spur formula 1.3.4. Three species in Europe, with one in Britain.

## *Hydatophylax infumatus* (McLachlan, 1865)

Listed as *Stenophylax infumatus* in Mosely (1939). Fore wing length: ♂♀ 15-17 mm (Fig. 680); uniformly dark in life. Very local in England, Wales and mainland Scotland; present in Ireland; streams and rivers, occasionally lakes. A secretive species, rarely encountered. Throughout much of Europe except extreme south. Flight period: June-July. ♂ genitalia as in Fig. 681; ♀ genitalia as in Fig. 682.

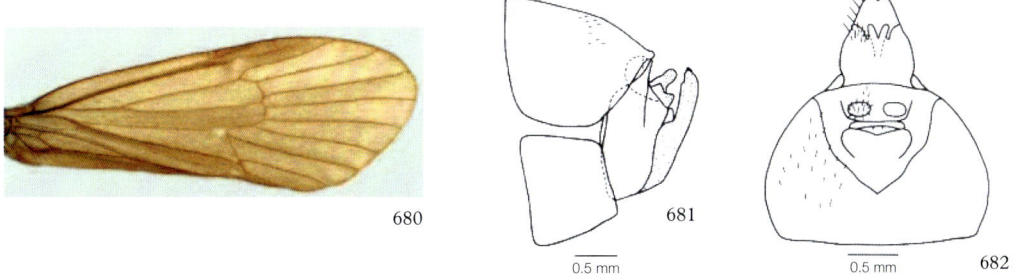

Figures 680-682. *Hydatophylax infumatus.* 680 fore wing pattern; 681 male genitalia lateral, aedeagus omitted; 682 female genitalia ventral

# Family SERICOSTOMATIDAE (2 genera, 2 species)

Fore wing with forks 1, 2, 3 and 5 present, hind wing with forks 1, 2 and 5; discoidal cell closed in fore wing, open or closed in hind wing (Figs 683, 684). Spur formula 2.2.4. Ocelli absent. Male maxillary palps highly modified.

## Key to genera of Sericostomatidae

1. Fore wing reddish-brown in life; in hind wing discoidal cell open (Fig. 683) ....................
..................................................................................... *Sericostoma* (p. 150)

- Fore wing blackish in life; in hind wing discoidal cell closed (Fig. 684) ... *Notidobia* (p. 151)

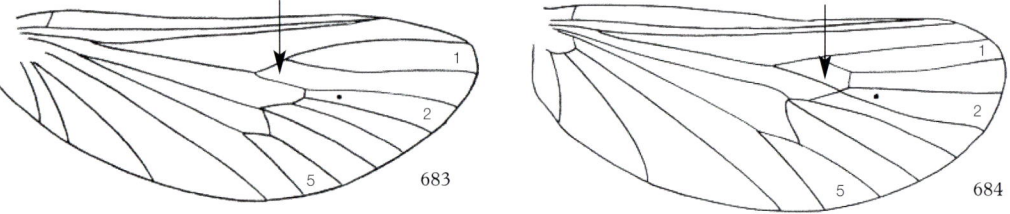

683     684

## Genus SERICOSTOMA Latreille, 1825

A large and complex genus in Europe with about 30 named species but the range of specific variation needs more examination. There is currently intensive study being undertaken on *Sericostoma personatum* and *S. schneideri* (listed as *S. flavicorne* in Malicky, 2004). These may be separate morphological species with distinct geographical distributions, or they may represent ecological forms, because there is some evidence that one group is confined to lowland rivers, with the other at higher altitudes. The two forms can apparently be separated more easily as larvae than as adults, and both larval colour patterns have been collected in Britain (Andrew Godfrey, pers. comm.). It is therefore possible that this species may eventually be split into two, with both species present in Britain.

### *Sericostoma personatum* (Spence, 1826)

This is the angler's Welshman's Button. Fore wing length: ♂ 9-14 mm, ♀ 10-16 mm (Figs 685-687); reddish brown in life. Common throughout most of Britain; present in Ireland; rivers, streams and stony lakes. Throughout most of Europe. Flight period: May-September; usually on the wing much later than *Notidobia*. ♂ genitalia as in Fig. 688; ♀ genitalia as in Fig. 689.

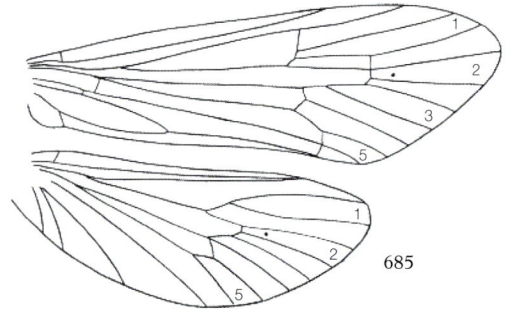

685

686

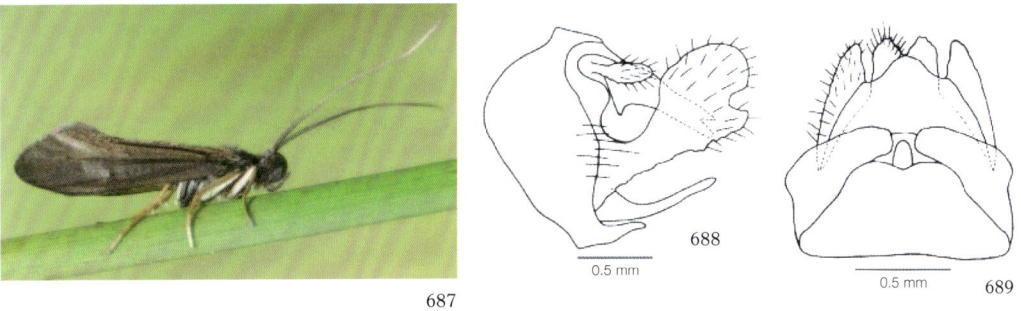

Figures 685-689. *Sericostoma personatum*. 685 wing venation; 686 wing pattern;
687 live specimen; 688 male genitalia lateral, aedeagus omitted; 689 female genitalia ventral

# Genus NOTIDOBIA Stephens, 1829

Eight European species, with just one in Britain.

## *Notidobia ciliaris* (Linnaeus, 1761)

Fore wing length: ♂♀ 9-11 mm (Figs 690, 691); blackish in life, always darker than *Sericostoma*. Not common, England as far north as Cheshire and Nottinghamshire and Wales; no records from Ireland; slow rivers and canals, and large drainage dykes. Throughout most of Europe. Flight period: May-June. ♂ genitalia as in Fig. 692; ♀ genitalia as in Fig. 693.

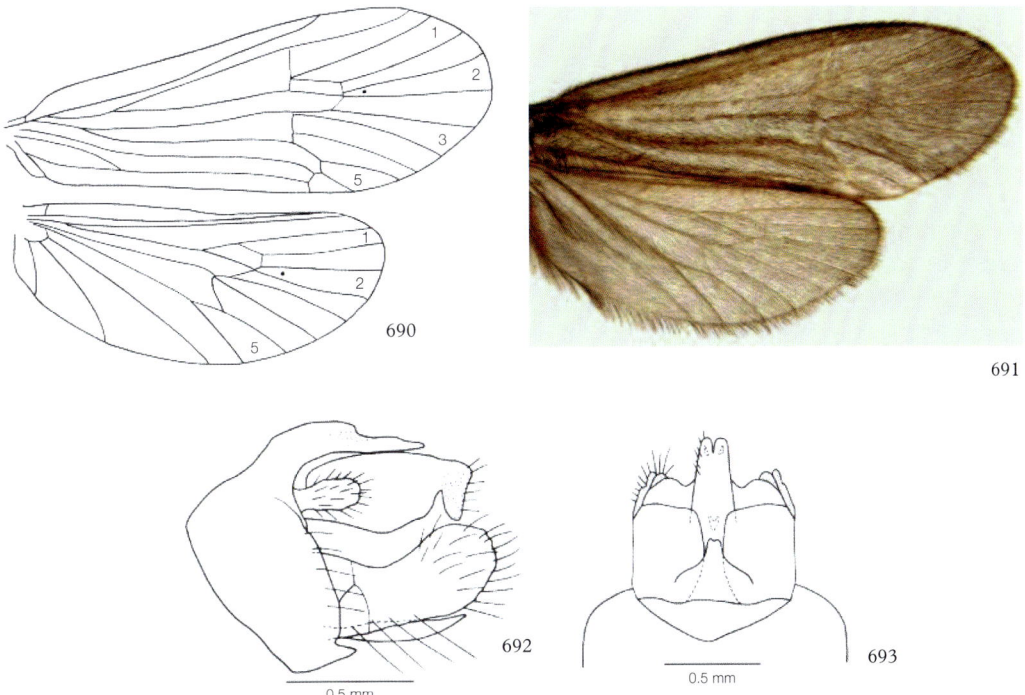

Figures 690-693. *Notidobia ciliaris*. 690 wing venation; 691 wing pattern;
692 male genitalia lateral, aedeagus omitted; 693 female genitalia ventral

# Family BERAEIDAE (3 genera, 4 species)

Wing venation highly modified in both sexes, with sexual dimorphism in all three genera. Spur formula 2.2.4. Ocelli absent. All species have black wings. A key to the three genera of Beraeidae is not very practicable and the four British species are easily recognised from the genitalia drawings.

## Genus BERAEA Stephens, 1833

Of nearly 20 European species just two occur in Britain.

### *Beraea maurus* (Curtis, 1834)

Fore wing length: ♂♀ 4-5 mm (Figs 694-698); male fore wing with oval pouch of scent scales at base (Figs 694, 696). Throughout much of Britain though few records from SE. England; present in Ireland; springs and trickles, including those in woodland. Throughout much of Europe except east. Flight period: May-August. ♂ genitalia as in Fig. 699; ♀ genitalia as in Fig. 700.

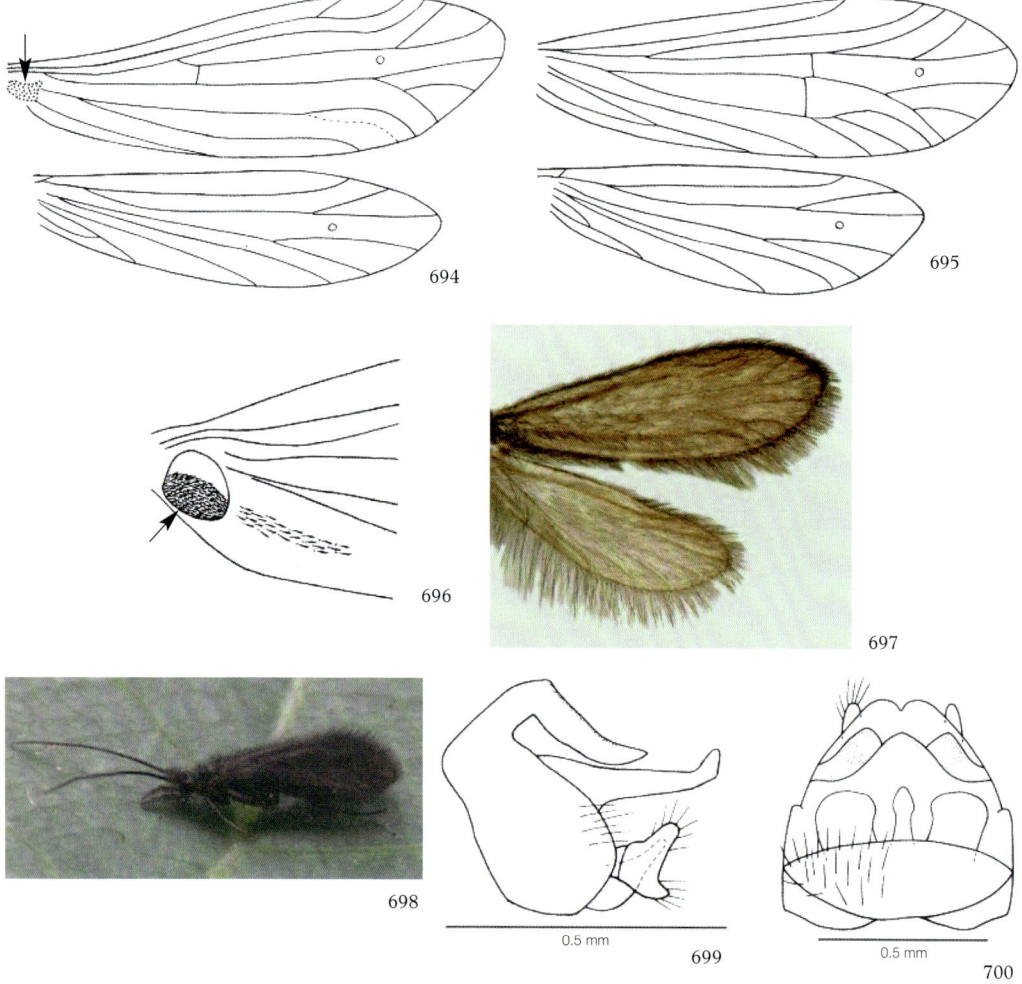

Figures 694-700. *Beraea maurus.* 694 male wing venation; 695 female wing venation; 696 male wing pouch; 697 wing pattern; 698 live specimen; 699 male genitalia lateral, aedeagus omitted; 700 female genitalia ventral

## *Beraea pullata* (Curtis, 1834)

Fore wing length: ♂♀ 4-6 mm (Figs 701-704); male fore wing with pouch of scent scales differently shaped from *B. maurus* (Fig. 703). Common throughout England, Wales and mainland Scotland; present in Ireland; flowing marshes, springs and margins of streams, preferring open rather than shaded sites. Throughout much of Europe. Flight period: May-July. ♂ genitalia as in Fig. 705; ♀ genitalia as in Fig. 706.

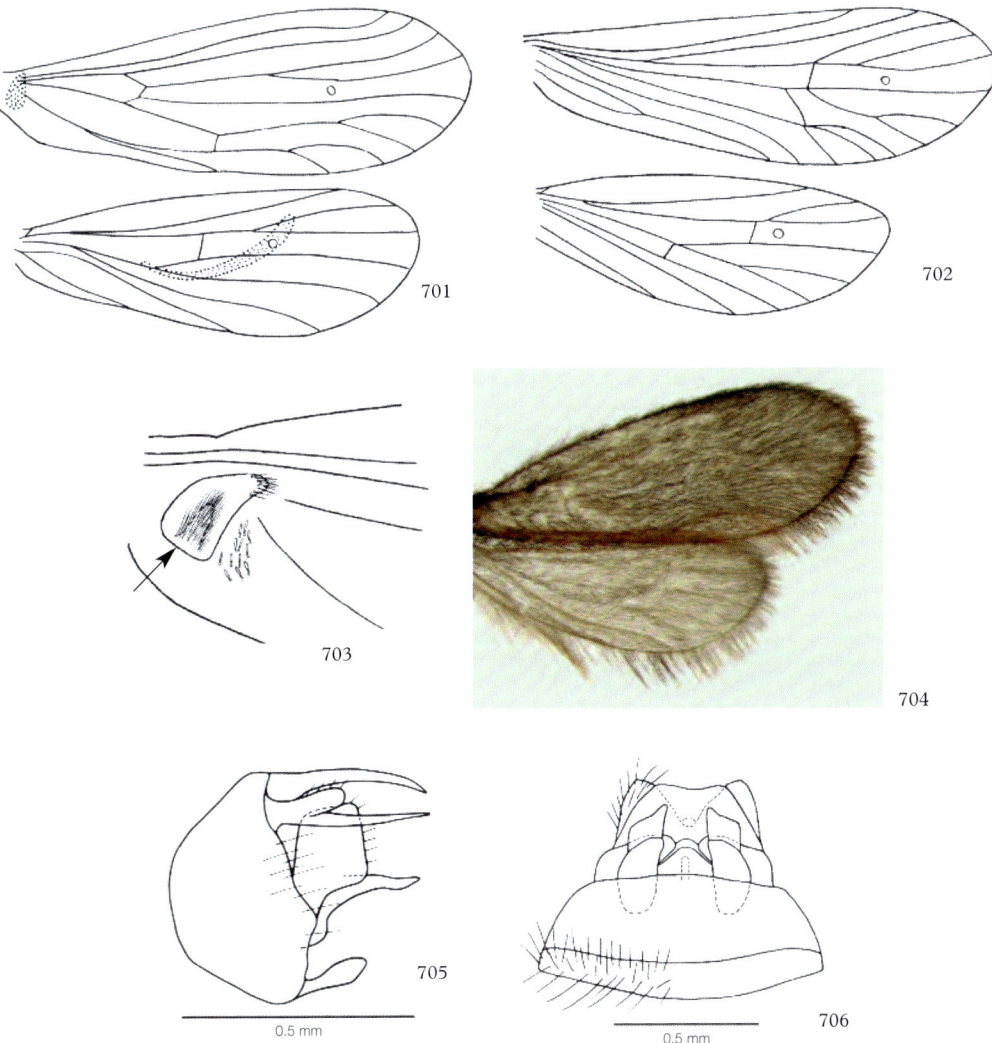

Figures 701-706. *Beraea pullata*. 701 male wing venation; 702 female wing venation; 703 male wing pouch; 704 wing pattern; 705 male genitalia lateral, aedeagus omitted; 706 female genitalia ventral

# Genus BERAEODES Eaton, 1867

*B. minutus* is the only European species.

## *Beraeodes minutus* (Linnaeus, 1761)

Listed as *Beraeodes minuta* in Mosely (1939). Fore wing length: ♂♀ 4-5 mm (Figs 707-709); wings narrower and more pointed than in the other members of the family; male fore wing with small fold containing scent scales near base (Fig. 707). Not common in England and Wales, few records from southern Scotland; present in Ireland; rivers, streams, edges of ponds and lakes, often associated with exposed roots of marginal plants. Throughout most of Europe. Flight period: May-June. ♂ genitalia as in Fig. 710; ♀ genitalia as in Fig. 711.

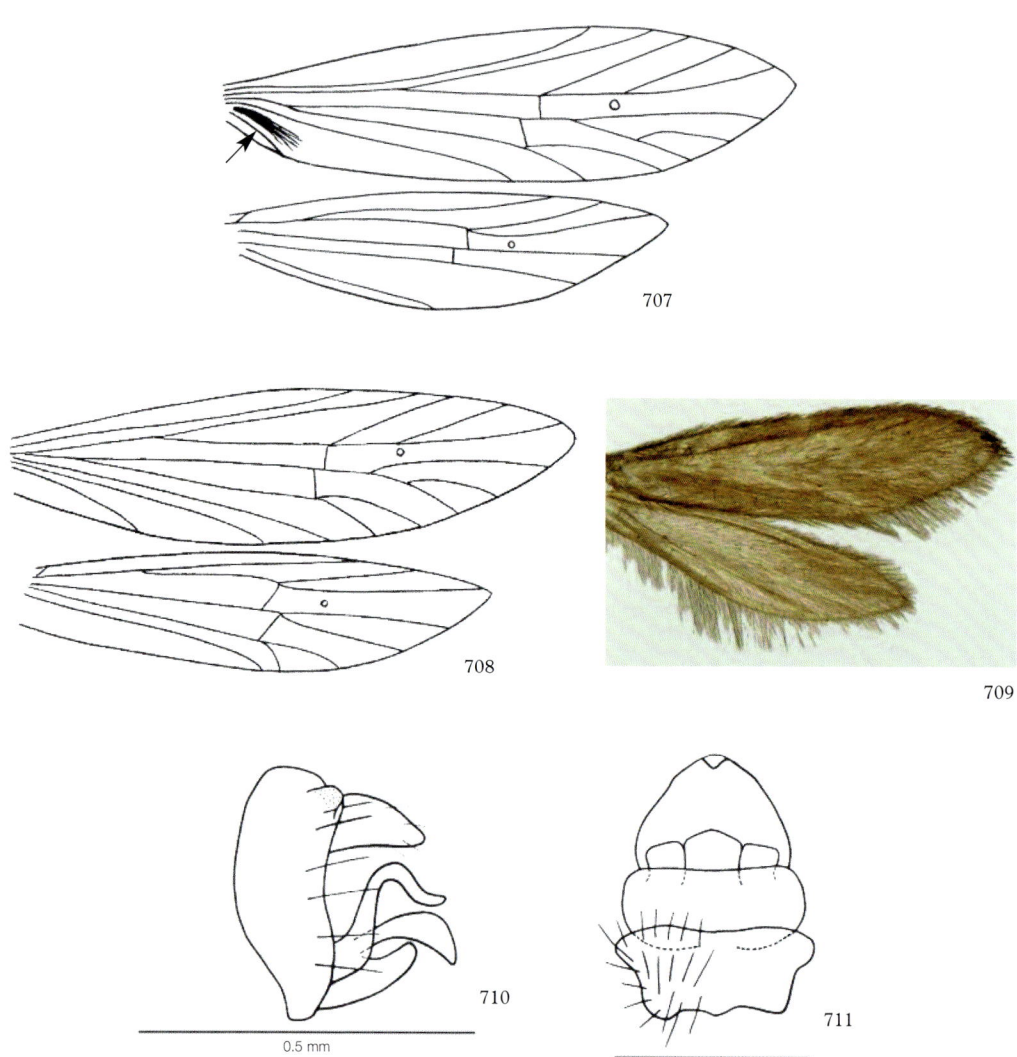

Figures 707-711. *Beraeodes minutus*. 707 male wing venation; 708 female wing venation; 709 wing pattern; 710 male genitalia lateral, aedeagus omitted; 711 female genitalia ventral

# Genus ERNODES Wallengren, 1891

About 15 European species with just one in Britain.

## *Ernodes articularis* (Pictet, 1834)

Fore wing length: ♂♀ 4-5 mm (Figs 712-714); no scent scales in male. Very local in parts of SW. England, Wales and adjacent English counties, and Derbyshire; no records from Ireland; calcareous streams and trickles. Mainly southern and central Europe. Flight period: May-July. ♂ genitalia as in Fig. 715; ♀ genitalia as in Fig. 716.

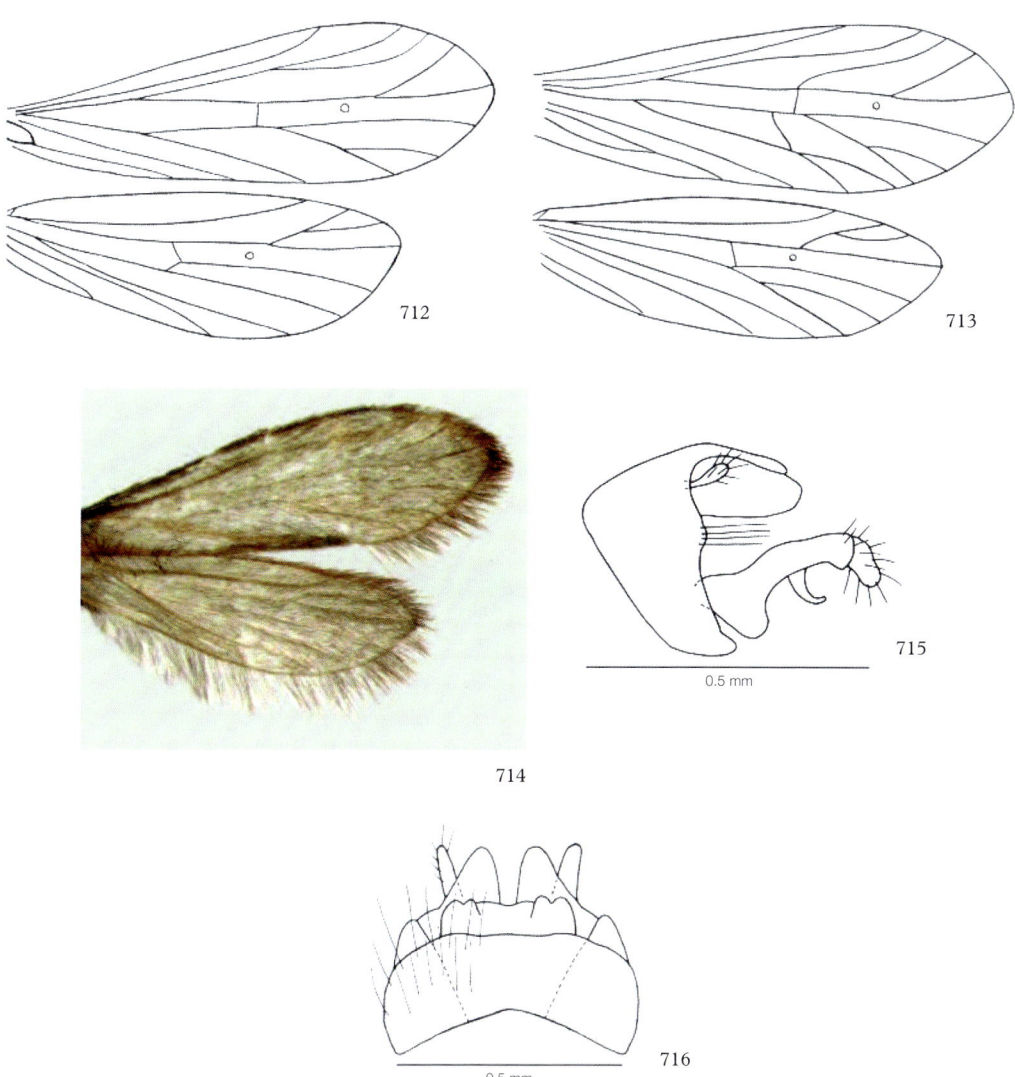

Figures 712-716. *Ernodes articularis.* 712 male wing venation; 713 female wing venation; 714 wing pattern; 715 male genitalia lateral, aedeagus omitted; 716 female genitalia ventral

# Family ODONTOCERIDAE (1 genus, 1 species)

Wing venation highly modified in male, forks 1 and 2 apparently present in both wings (Fig. 717); female with forks 1, 2, 3 and 5 in both wings (Fig. 718); discoidal cell closed in both wings. Spur formula 2.4.4. Ocelli absent. *O. albicorne* is the only British species, with two more found in Europe.

## Genus ODONTOCERUM Leach, 1815

### *Odontocerum albicorne* (Scopoli, 1763)

This is the angler's Silver or Grey Sedge. Fore wing length: ♂ 12-16 mm, ♀ 14-18 mm (Figs 717-720); the narrow wings are not always as strongly marked as in Fig. 720. The toothed antennae are very distinctive (Fig. 721) when viewed from the correct angle. Common throughout England, Wales and mainland Scotland; present in Ireland; stony streams and rivers. Throughout Europe, less common in north. Flight period: May-September (October). ♂ genitalia as in Fig. 722; ♀ genitalia as in Fig. 723.

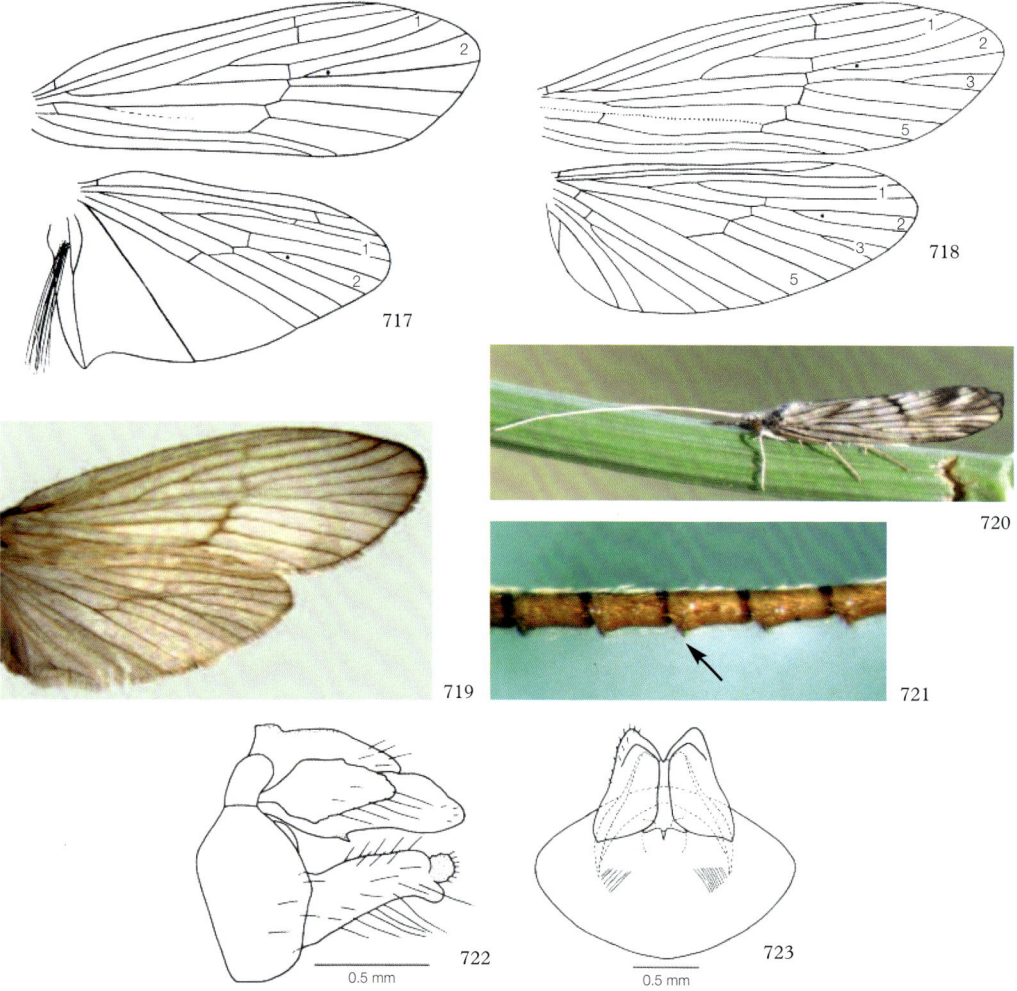

0.5 mm

0.5 mm

Figures 717-723. *Odontocerum albicorne*. 717 male wing venation; 718 female wing venation; 719 wing pattern; 720 live specimen; 721 closeup of antenna; 722 male genitalia lateral, aedeagus omitted; 723 female genitalia ventral

# Family MOLANNIDAE (1 genus, 2 species)

Wing venation highly modified in both sexes, with homology of veins unclear (Figs 728, 729). Spur formula 2.4.4. Ocelli absent.

## Genus MOLANNA Curtis, 1834

Two of the four European species are found in Britain.

### *Molanna albicans* (Zetterstedt, 1840)
= *palpata* McLachlan, 1877

Listed as *M. palpata* in Mosely (1939) and Macan (1973). Fore wing length: ♂ 8-10 mm, ♀ 12-14 mm (Fig. 724); wings greyish and partly rolled round the body at rest. Local in Wales, Yorkshire and one isolated site in northern England, but surprisingly absent from Lake District, also local in central and northern Scotland; present in Ireland; mainly upland lakes. Mainly central and northern Europe. Flight period: June-September. ♂ genitalia as in Figs 725, 726; ♀ genitalia as in Fig. 727.

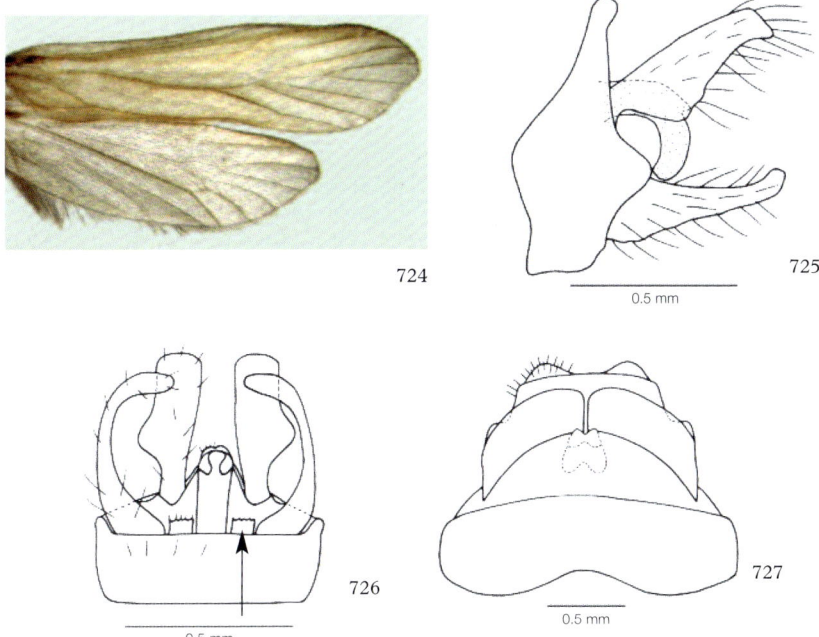

Figures 724-727. *Molanna albicans.* 724 wing pattern; 725 male genitalia lateral, aedeagus omitted; 726 male genitalia ventral; 727 female genitalia ventral

## *Molanna angustata* **Curtis, 1834**

Fore wing length: ♂ 9-12 mm, ♀ 11-15 mm (Figs 728-731); similar to *M. albicans* but wings usually more tightly rolled. Common in England, a few records from Wales; no records from Ireland; ponds, lakes, canals and slow rivers. Mainly central and northern Europe. Flight period: May-September. ♂ genitalia as in Figs 732, 733; ♀ genitalia as in Fig. 734.

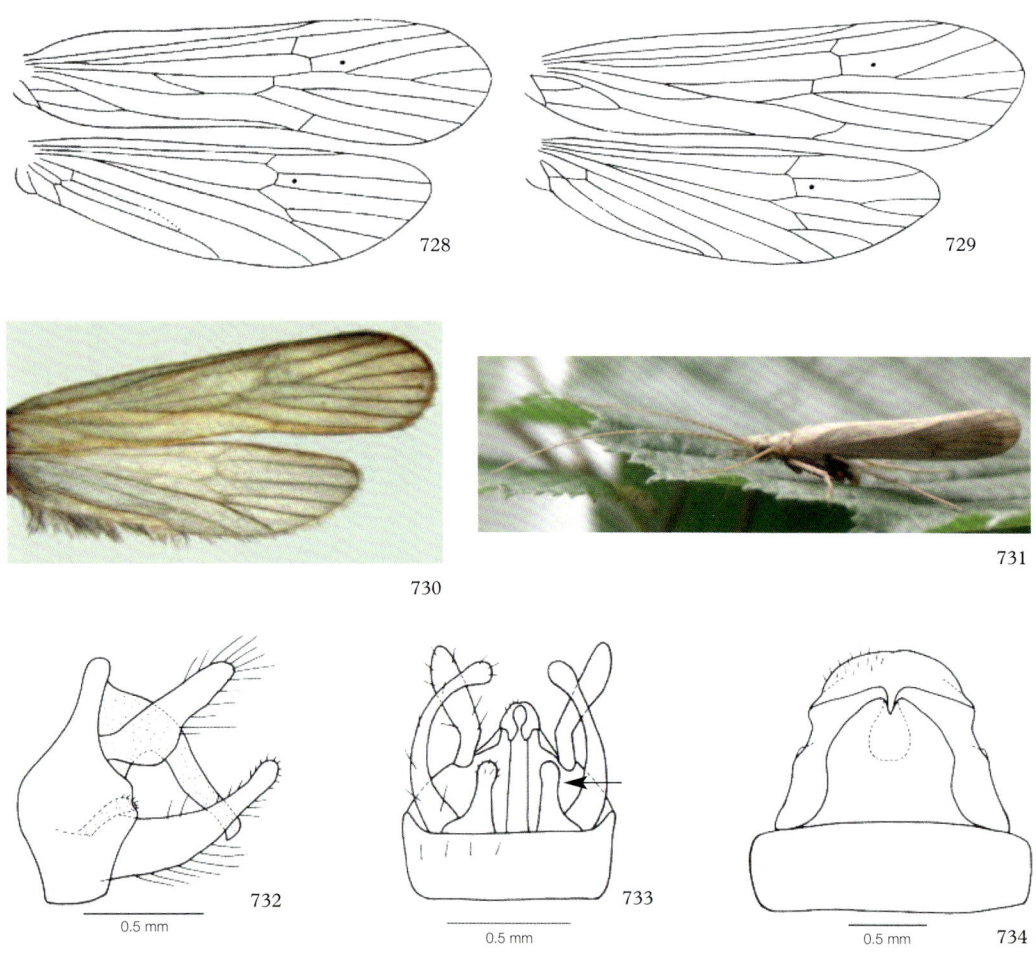

Figures 728-734. *Molanna angustata*. 728 male wing venation; 729 female wing venation; 730 wing pattern; 731 live specimen; 732 male genitalia lateral, aedeagus omitted; 733 male genitalia ventral; 734 female genitalia ventral

# Family LEPTOCERIDAE (10 genera, 31 species)

Wing venation differs in each genus but discoidal cell always closed in fore wing, open in hind wing. Spur formula 0.2.2, 1.2.2 or 2.2.2. Ocelli absent. Members of this family can usually be easily recognised by the very long antennae (at least twice the length of the fore wing); this is the only family with just 2 spurs on the hind tibia. A predominantly day-flying group, often with complex swarming patterns, though some species are also attracted to light.

## Key to genera or generic groups of Leptoceridae

1. In fore wing basal section of vein M absent (Fig. 735) ....................................................
.................................................................... *Triaenodes* (p. 162) and *Ylodes* (p. 163)

- In fore wing basal section of vein M present, though occasionally represented only by a transparent line (Fig. 736) ................................................................................................ 2

735

736

2. In fore wing M runs straight to wing margin without forking (Fig. 737) ..... *Oecetis* (p. 178)

- In fore wing M forked (Fig. 738) ................................................................................ 3

737

738

3. Spur formula 2.2.2 ............................................ *Athripsodes* (p. 164) and *Ceraclea* (p. 168)

- Spur formula 0.2.2 or 1.2.2 ......................................................................................... 4

4. Spur formula 1.2.2 ................................................ *Erotesis* (p. 160) and *Adicella* (p. 160)

- Spur formula 0.2.2 ..................................................................................................... 5

5. In fore wing apical fork 1 sessile (Fig. 739) ........................................ *Mystacides* (p. 174)

- In fore wing apical fork 1 stalked (Fig. 740) ............................................................... 6

739

740

6. In fore wing apical fork 4 sessile (Fig. 741) ........................................ *Leptocerus* (p. 172)

- In fore wing apical fork 4 stalked (Fig. 742) ........................................... *Setodes* (p. 176)

741

742

# Genus EROTESIS McLachlan, 1877

Fore wing with forks 1 and 5 present, hind wing with fork 1 only (Fig. 743). Spur formula 1.2.2. Four species in Europe with just one in Britain.

## *Erotesis baltica* **McLachlan, 1877**

Fore wing length: ♂♀ 7-8 mm (Figs 743, 744); greyish brown. Lake District, Anglesey, East Anglia and Hampshire, not common; present in Ireland; permanently wet fen and associated dykes and lake margins. Throughout much of Europe except south. Flight period: June-September. ♂ genitalia as in Fig. 745; ♀ genitalia as in Fig. 746.

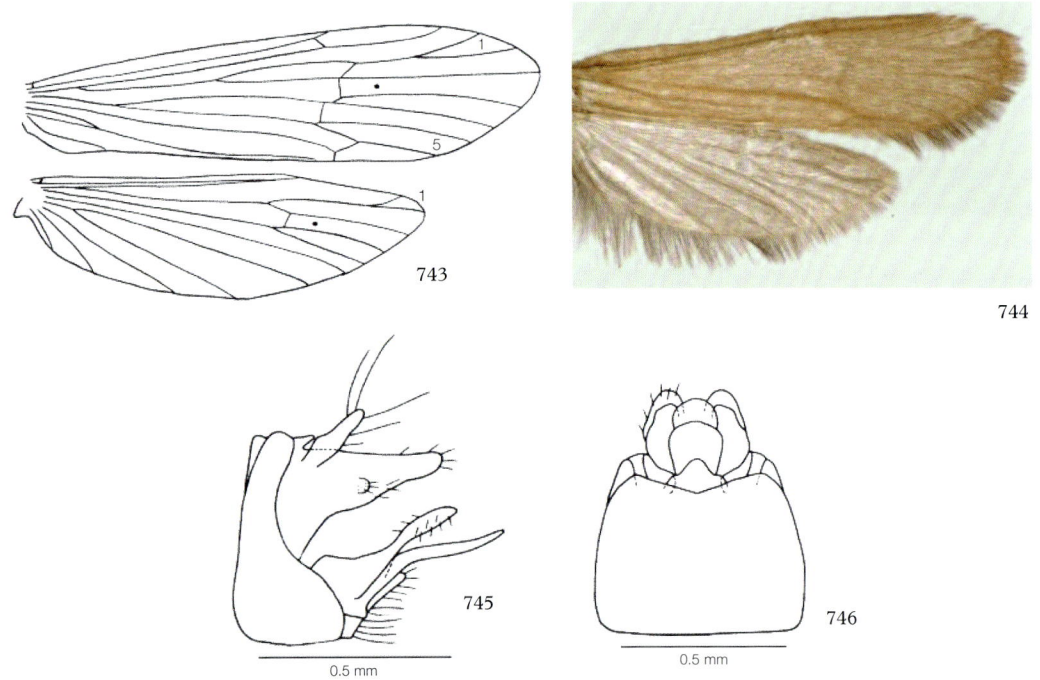

Figures 743-746. *Erotesis baltica*. 743 wing venation; 744 wing pattern; 745 male genitalia lateral, aedeagus omitted; 746 female genitalia ventral

# Genus ADICELLA McLachlan, 1877

Fore wing with forks 1 and 5 present, hind wing with fork 1 only (Fig. 750). Spur formula 1.2.2. Fifteen European species, with two in Britain.

## *Adicella filicornis* **(Pictet, 1834)**

Fore wing length: ♂ 6-8 mm, ♀ 6-7 mm (Fig. 747); dark reddish brown, almost black in life. Extremely local with only a few sites in Devon, Hampshire, Wales, Yorkshire and Lanarkshire; no records from Ireland; spring streams and trickles. Mainly southern and central Europe. Flight period: June-July. ♂ genitalia with narrow finger-like superior appendages (Fig. 748); ♀ genitalia with short hairs on terminal lobes (Fig. 749).

160

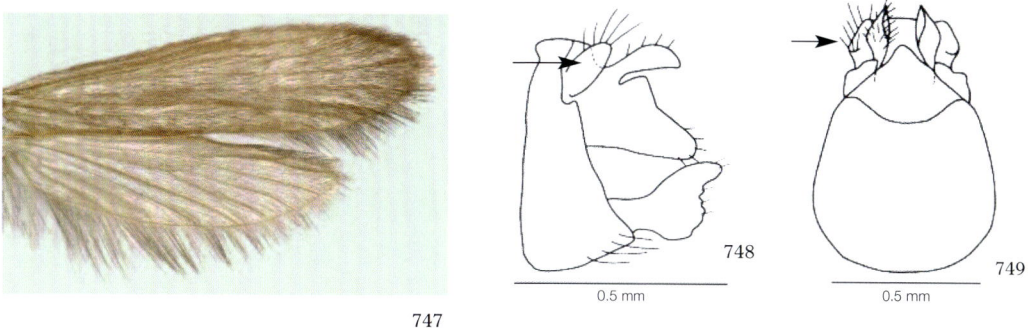

Figures 747-749. *Adicella filicornis*. 747 wing pattern;
748 male genitalia lateral, aedeagus omitted; 749 female genitalia ventral

## *Adicella reducta* (McLachlan, 1865)

Fore wing length: ♂ 5-8 mm, ♀ 6-7 mm (Figs 750-752); reddish brown. Common throughout England, Wales and mainland Scotland; present in Ireland; streams, flowing marshes, canals and rivers, often associated with tree roots and found in small impoverished water bodies, unlike most other leptocerids. Throughout much of Europe except east. Flight period: (May) June-August. ♂ genitalia with broad superior appendages (Fig. 753), ♀ genitalia with very long hairs on terminal lobes (Fig. 754).

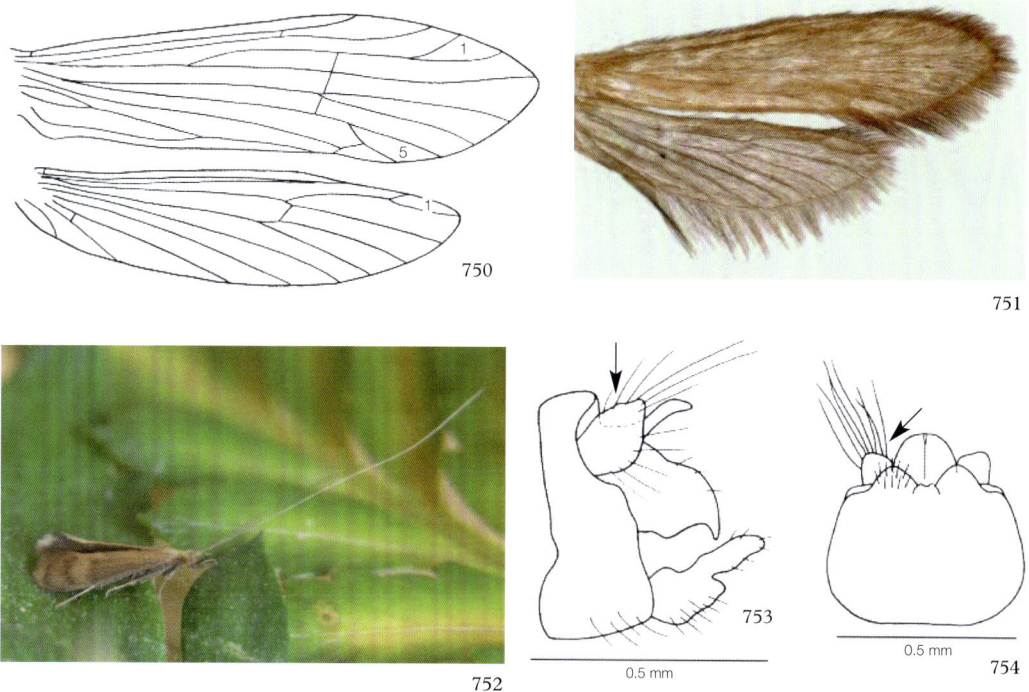

Figures 750-754. *Adicella reducta*. 750 wing venation; 751 wing pattern; 752 live specimen;
753 male genitalia lateral, aedeagus omitted; 754 female genitalia ventral

# Genus TRIAENODES McLachlan, 1865

Fore wing with forks 1, 2 and 5 present (though fork 2 may not be correctly identified), hind wing with only fork 1 (Fig. 755). Spur formula 1.2.2. Just one of the five European species occurs in Britain.

## *Triaenodes bicolor* (Curtis, 1834)

This is the angler's Bicolor Sedge. Fore wing length: ♂ 6-7 mm, ♀ 7-10 mm (Figs 755, 756); fore wing bright reddish brown, hind wing grey. Common throughout most of Britain; present in Ireland; ponds, lakes, slow rivers, canals and permanently wet fen. Throughout most of Europe. Flight period: June-September. ♂ genitalia as in Fig. 757; ♀ genitalia as in Fig. 758.

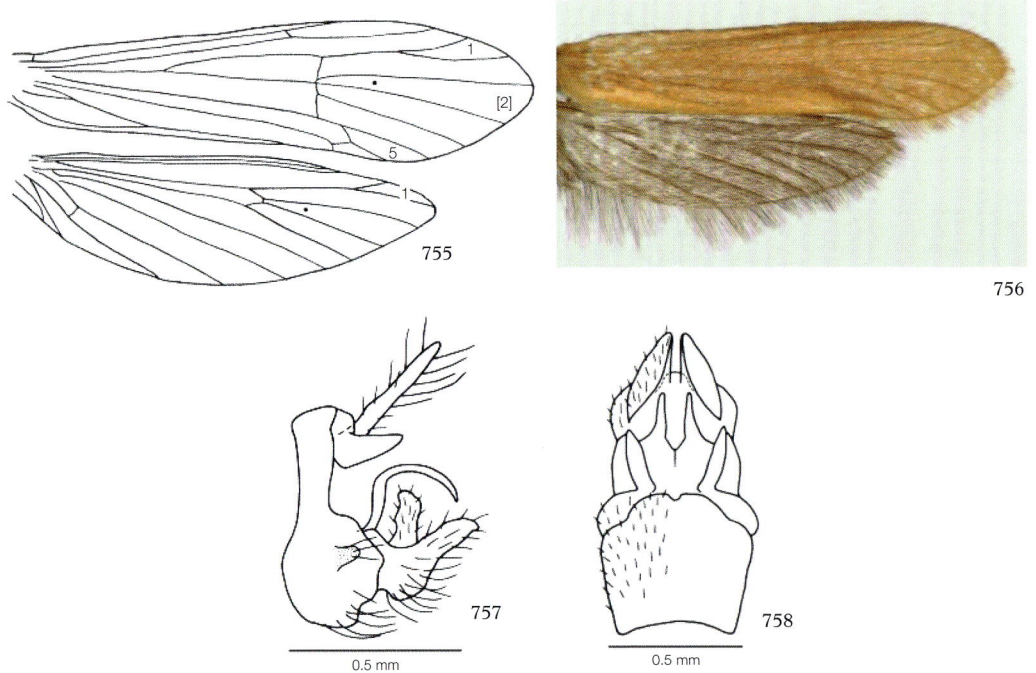

Figures 755-758. *Triaenodes bicolor*. 755 wing venation; 756 wing pattern; 757 male genitalia lateral, aedeagus omitted; 758 female genitalia ventral

# Genus YLODES Milne, 1934

Fore wing with forks 1, 2 and 5 present (though fork 2 may not be correctly identified), hind wing with only fork 1, as in *Triaenodes* (Fig. 755). Spur formula 1.2.2. Ten European species of which three occur in Britain.

## *Ylodes conspersus* (Rambur, 1842)

Listed as *Triaenodes conspersus* in Mosely (1939) and *T. conspersa* in Macan (1973). Fore wing length: ♂ 8-10 mm, ♀ 7-11 mm (Fig. 759); greyish hairs forming mottled pattern. S. England and Midlands, not common (early records from East Anglia may refer to *Y. reuteri*); no records from Ireland; medium to large rivers. Mainly southern and central Europe. Flight period: July-September. ♂ genitalia with pointed ventral lobe on claspers (Fig. 760); ♀ genitalia with V-shaped excision in vulvar scale (Fig. 761).

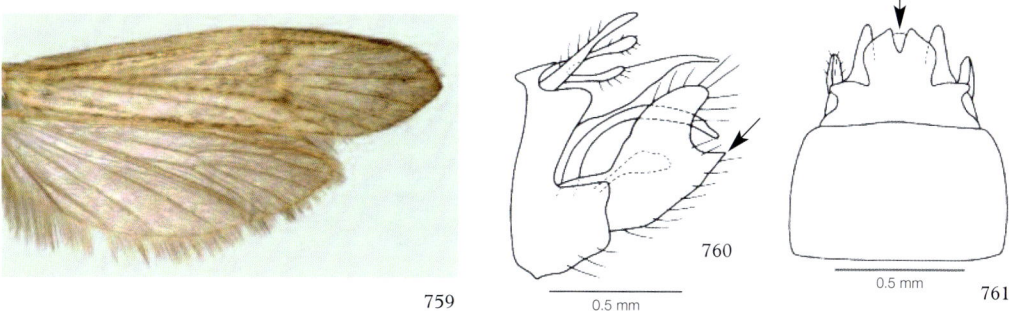

Figures 759-761. *Ylodes conspersus*. 759 wing pattern;
760 male genitalia lateral, aedeagus omitted; 761 female genitalia ventral

## *Ylodes reuteri* (McLachlan, 1880)
= *simulans*; misidentified by some authors

Listed as *Triaenodes reuteri* in Macan (1973); not recognised until Kimmins (1964) [misidentified]; see Pelham-Clinton (1966b). Fore wing length: ♂ 6-9 mm, ♀ 8-9 mm (Fig. 762); coloration similar to *Y. conspersus*. S. England, East Anglia, Yorkshire and Orkney, not common; present in Ireland; brackish water, especially coastal marshes. Mainly central Europe. Flight period: July. ♂ genitalia with very small ventral lobe on claspers (Fig. 763); ♀ genitalia with short and inconspicuous segment X (Fig. 764).

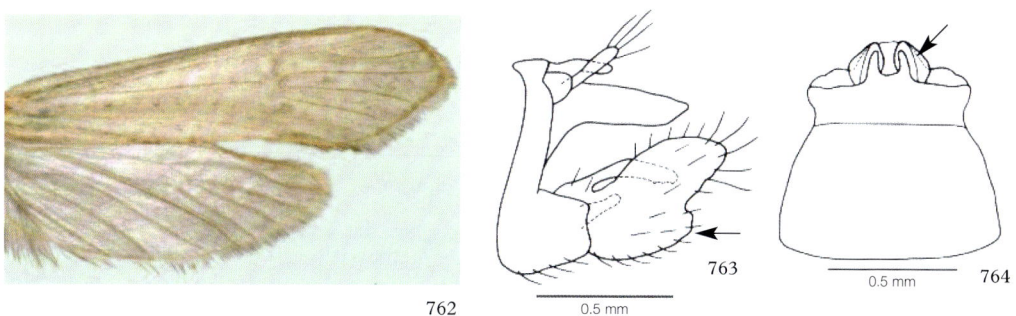

Figures 762-764. *Ylodes reuteri*. 762 wing pattern;
763 male genitalia lateral, aedeagus omitted; 764 female genitalia ventral

## *Ylodes simulans* (Tjeder, 1929)

Listed as *Triaenodes simulans* in Macan (1973). This species was first recorded in Britain by Morton (1906) under the name *Triaenodes reuteri*, an error later corrected by Morton (1931). However, Morton's records were apparently overlooked by Mosely (1939) and the species was re-introduced to the British list by Kimmins (1949). The later confusion between the two species by Kimmins (1964) was finally resolved by Pelham-Clinton (1966b). Fore wing length: ♂♀ 6-8 mm (Fig. 765); coloration similar to *Y. conspersus*. SW. England, Wales and some adjoining English counties with one record from central Scotland, not common; no records from Ireland; rivers. Mainly central and northern Europe. Flight period: July. ♂ genitalia with well-defined ventral lobe on claspers, almost as large as dorsal lobe (Fig. 766); ♀ genitalia with prominent segment X (Fig. 767).

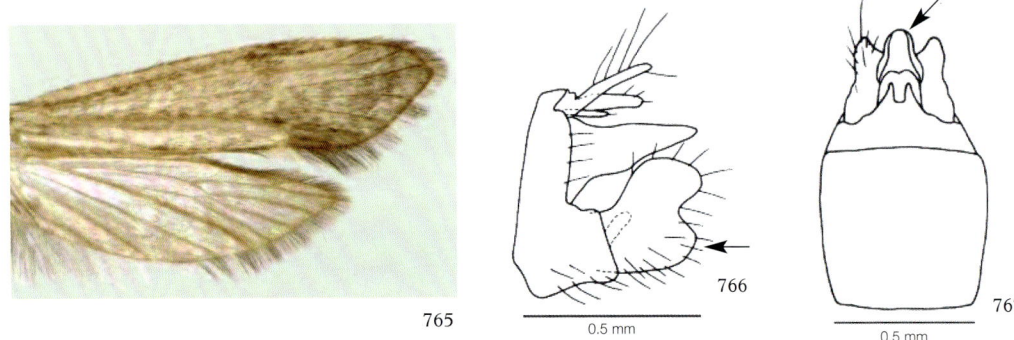

Figures 765-767. *Ylodes simulans*. 765 wing pattern;
766 male genitalia lateral, aedeagus omitted; 767 female genitalia ventral

# Genus ATHRIPSODES Billberg, 1820

Fore wing with forks 1 and 5 in male, 1, 3 and 5 in female; hind wing with fork 1 and 5 in both sexes (Fig. 784). Spur formula 2.2.2 (spurs on fore tibia very small). About 25 species in Europe with five in Britain. Together with the genus *Ceraclea* these are the angler's Brown or Black Silverhorns. The sexual dimorphism in wing venation also occurs in *Ceraclea*, emphasising the close association of these two genera, as this dimorphism does not occur elsewhere in the family.

## *Athripsodes albifrons* (Linnaeus, 1758)
= form *interjectus* (McLachlan, 1881)

Listed as *Leptocerus albifrons* in Mosely (1939), and with *L. interjectus* as a separate species. Fore wing length: ♂ 7-9 mm, ♀ 6-8 mm (Figs 768, 769); brown with four white markings. The form *interjectus* has black wings in life, though these fade to brown after death, and a similar colour form also occurs in *A. commutatus*. The head has a conspicuous patch of white hairs. Common in England, Wales and mainland Scotland, less common in the north; present in Ireland; rivers and large streams. Throughout Europe. Flight period: May-September. ♂ genitalia as in Fig. 770; ♀ genitalia as in Fig. 771.

Figures 768-771. *Athripsodes albifrons*. 768 wing pattern; 769 live specimen;
770 male genitalia lateral, aedeagus omitted; 771 female genitalia ventral

## *Athripsodes bilineatus* (Linnaeus, 1758)

Listed as *Leptocerus bilineatus* in Mosely (1939). Fore wing length: ♂ 7-9 mm, ♀ 7-10 mm (Figs 772, 773); shiny black with white markings similar to the *interjectus* form of *A. albifrons*, but the head is black, not white. Throughout England, Wales and mainland Scotland, but not common in SE. England; present in Ireland; streams, rivers, lakes. Throughout much of Europe. Flight period: May-August (September). ♂ genitalia as in Fig. 774; ♀ genitalia as in Fig. 775.

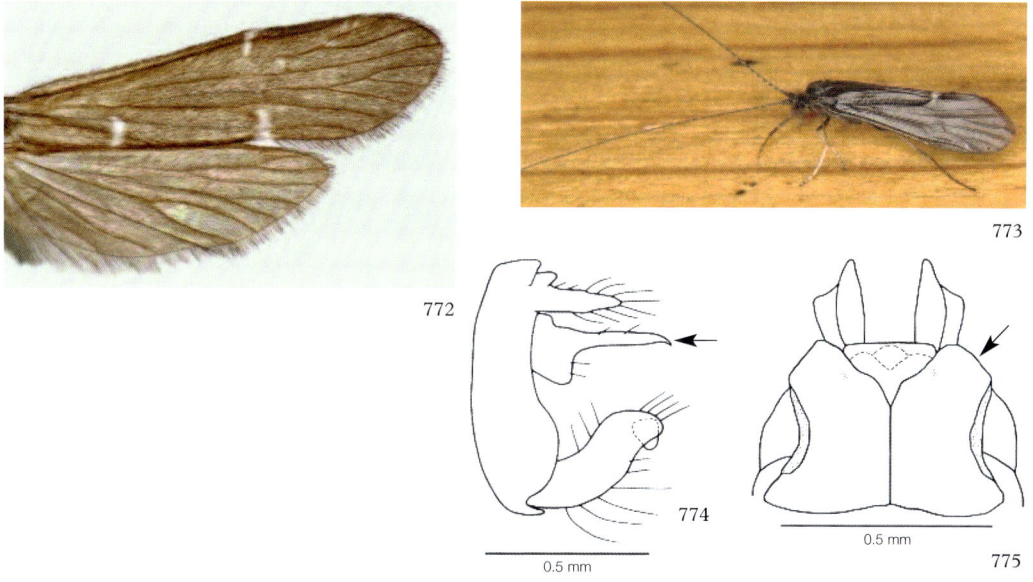

Figures 772-775. *Athripsodes bilineatus*. 772 wing pattern; 773 live specimen [photo: Peter Barnard];
774 male genitalia lateral, aedeagus omitted; 775 female genitalia ventral

165

## *Athripsodes commutatus* (Rostock, 1874)

Listed as *Leptocerus commutatus* in Mosely (1939). Fore wing length: ♂ 7-10 mm, ♀ 7-8 mm (Fig. 776); similar markings to *A. albifrons* and *A. bilineatus*, but cream-coloured rather than white. The terminal section of the antennae is white, with no black bands, and head has yellowish hairs. Very local in SW. England, Wales and bordering English counties, more widespread in N. England and Scotland; present in Ireland; large stony streams and rivers. Throughout much of Europe except south. Flight period: July-August. ♂ genitalia as in Figs 777, 778; ♀ genitalia as in Fig. 779.

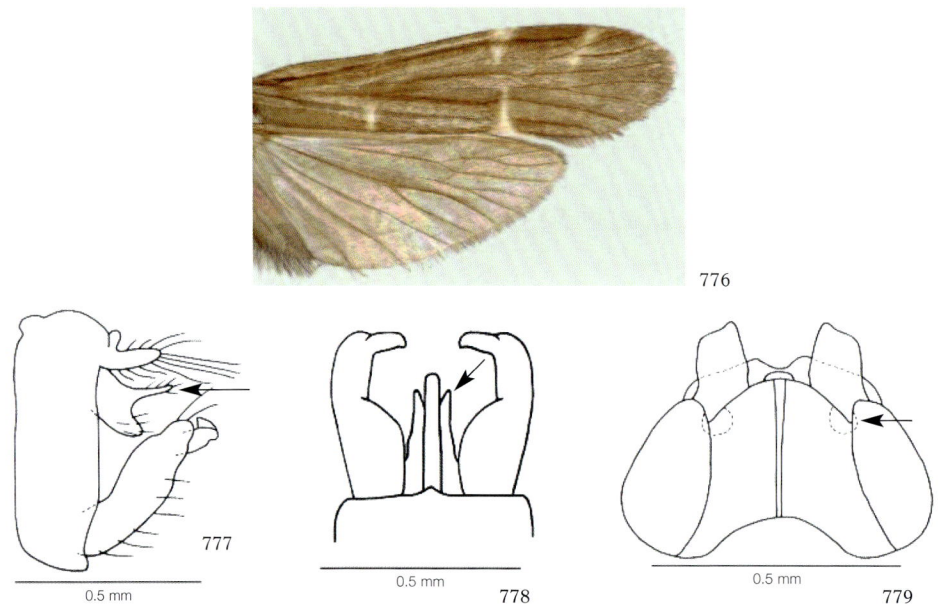

Figures 776-779. *Athripsodes commutatus.* 776 wing pattern; 777 male genitalia lateral, aedeagus omitted; 778 male genitalia ventral, aedeagus omitted; 779 female genitalia ventral

## *Athripsodes cinereus* (Curtis, 1834)

Listed as *Leptocerus cinereus* in Mosely (1939). Fore wing length: ♂ 9-11 mm, ♀ 8-10 mm (Figs 780, 781); a large brown species with variable markings; the main stem of vein M appears as a pale line against the darker membrane. Common throughout Britain; present in Ireland; streams rivers, canals and lakes. Throughout much of Europe except extreme south. Flight period: (May) June-August. ♂ genitalia as in Fig. 782; ♀ genitalia as in Fig. 783.

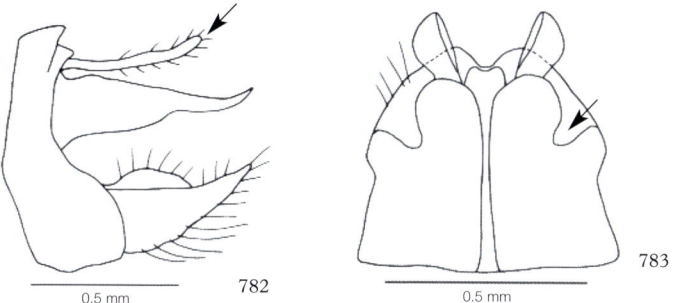

Figures 780-783. *Athripsodes cinereus*. 780 wing pattern; 781 live specimen [photo: Peter Barnard]; 782 male genitalia lateral, aedeagus omitted; 783 female genitalia ventral

## *Athripsodes aterrimus* (Stephens, 1836)

Listed as *Leptocerus aterrimus* in Mosely (1939). Fore wing length: ♂ 8-10 mm, ♀ 7-10 mm (Figs 784, 785); various shades of brown or almost black. Common throughout most of Britain; present in Ireland; ponds, lakes, slow rivers, lake outlets. Throughout much of Europe except extreme south. Flight period: May-September (October). ♂ genitalia as in Fig. 786; ♀ genitalia as in Fig. 787.

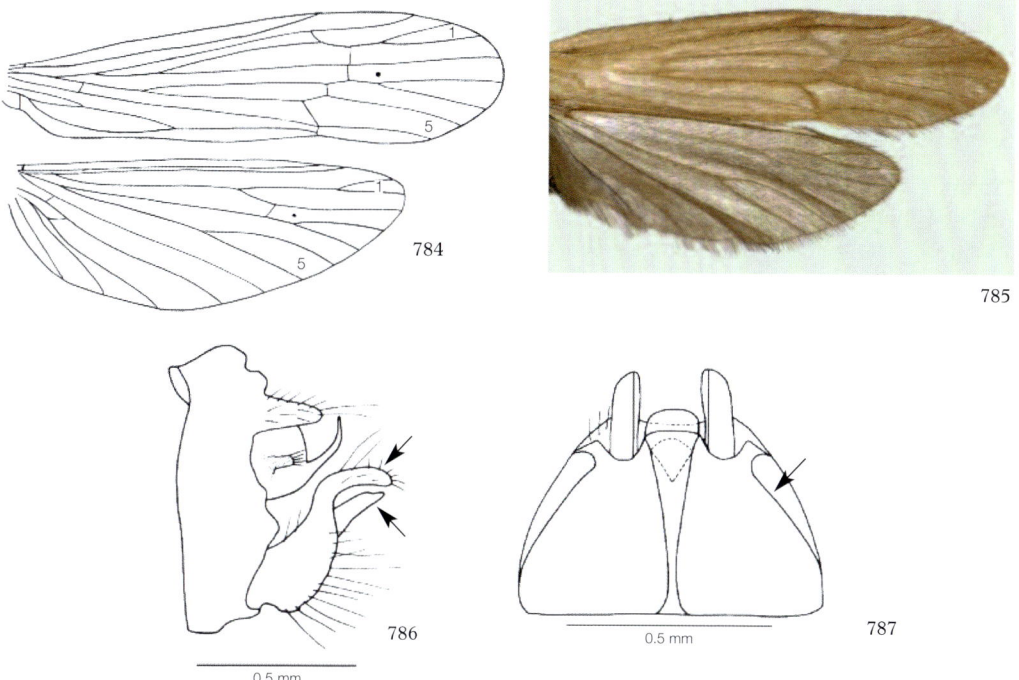

Figures 784-787. *Athripsodes aterrimus*. 784 wing venation; 785 wing pattern; 786 male genitalia lateral, aedeagus omitted; 787 female genitalia ventral

# Genus CERACLEA Stephens, 1829

Fore wing with forks 1 and 5 in male, 1, 3 and 5 in female; hind wing with fork 1 and 5 in both sexes (Figs 799, 800). Spur formula 2.2.2 (spurs on fore tibia very small). About 15 species in Europe of which six occur in Britain. Sexual dimorphism in wing venation, as in *Athripsodes*. Together with the genus *Athripsodes* these are the angler's Brown or Black Silverhorns. Although the genus was not formally separated from *Athripsodes* until Morse & Wallace (1976), the two groups of species (then within *Leptocerus*) were distinguished by earlier authors such as Mosely (1939), by the form of the maxillary palps. In what is now the genus *Ceraclea* both the 4th and 5th segments of the maxillary palps are subdivided into numerous small sclerites (Mosely described the segment as "flexible") (Fig. 789). In *Athripsodes* only the 5th segment is so divided (Fig. 788). This distinction can only be seen in fresh specimens or those preserved in fluid, so is not always of practical value.

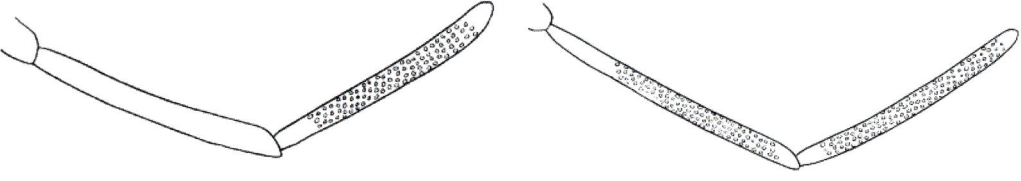

Figure 788. *Athripsodes* maxillary palp          Figure 789. *Ceraclea* maxillary palp

## *Ceraclea albimacula* (Rambur, 1842)
= *alboguttatus* (Hagen, 1860)

Listed as *Leptocerus alboguttatus* in Mosely (1939) and *Arthripsodes* [sic] *alboguttatus* in Macan (1973). *C. alboguttata* and *C. albimacula* have variously been regarded as separate species or synonyms by different authors. Malicky (2005) decided that, although the two forms appear distinct, based on the form of the male claspers, these differences represent each end of a continuous variation across the range of the species. Fore wing length: ♂ 11-13 mm, ♀ 10-12 mm (Fig.790); rather uniform yellowish brown. Throughout Britain, but not common; present in Ireland; canals, rivers and large streams. Throughout much of Europe, except east. Flight period: June-September. ♂ genitalia as in Fig. 791; ♀ genitalia as in Fig. 792.

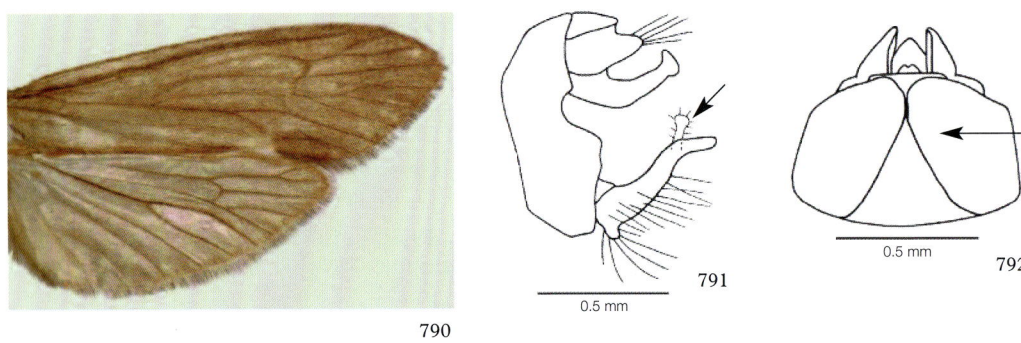

Figures 790-792. *Ceraclea albimacula*. 790 wing pattern;
791 male genitalia lateral, aedeagus omitted; 792 female genitalia ventral

## *Ceraclea fulva* (Rambur, 1842)

Listed as *Leptocerus fulvus* in Mosely (1939) and *Arthripsodes* [sic] *fulvus* in Macan (1973). Fore wing length: ♂ 10-13 mm, ♀ 10-12 mm (Fig. 793); similar to *C. albimacula*. Common throughout Britain, but less common in S. England; present in Ireland; lakes and slow rivers. Throughout much of Europe except south. Flight period: June-August (September). ♂ genitalia as in Fig. 794; ♀ genitalia as in Fig. 795.

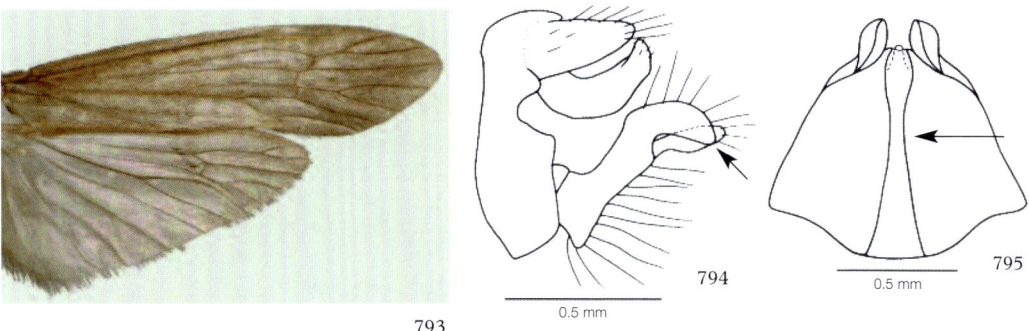

Figures 793-795. *Ceraclea fulva*. 793 wing pattern;
794 male genitalia lateral, aedeagus omitted; 795 female genitalia ventral

## *Ceraclea annulicornis* (Stephens, 1836)

Listed as *Leptocerus annulicornis* in Mosely (1939) and *Arthripsodes* [sic] *annulicornis* in Macan (1973). Fore wing length: ♂ 10-12 mm, ♀ 8-10 mm (Fig.796); similar to *C. albimacula*. Local throughout Britain, but absent from many areas; present in Ireland; stony rivers, occasionally lakes. Throughout much of Europe except south. Flight period: May-August. ♂ genitalia as in Fig. 797; ♀ genitalia as in Fig. 798.

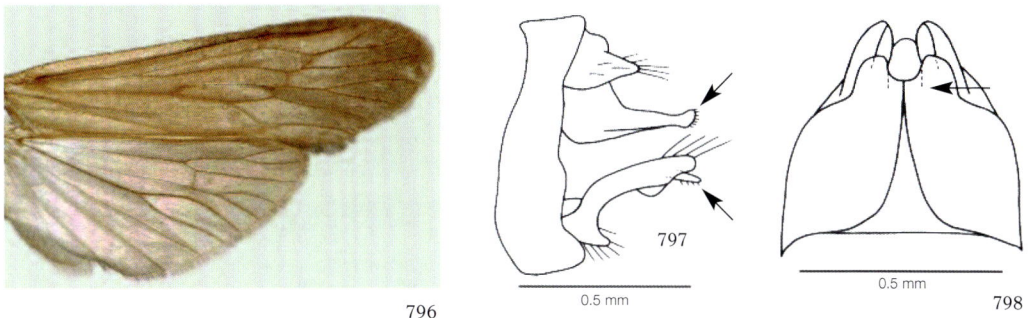

Figures 796-798. *Ceraclea annulicornis*. 796 wing pattern;
797 male genitalia lateral, aedeagus omitted; 798 female genitalia ventral

## *Ceraclea nigronervosa* (Retzius, 1783)

Listed as *Leptocerus nigronervosus* in Mosely (1939) and *Arthripsodes* [sic] *nigronervosus* in Macan (1973). Fore wing length: ♂ 8-10 mm ♀ 10-13 mm (Figs 799-801); a large dark species with conspicuously darker veins. Common throughout Britain; present in Ireland; rivers and lakes. Throughout much of Europe except south. Flight period: May-July; an early flight period especially in the southern half of Britain. ♂ genitalia as in Fig. 802; ♀ genitalia as in Fig. 803.

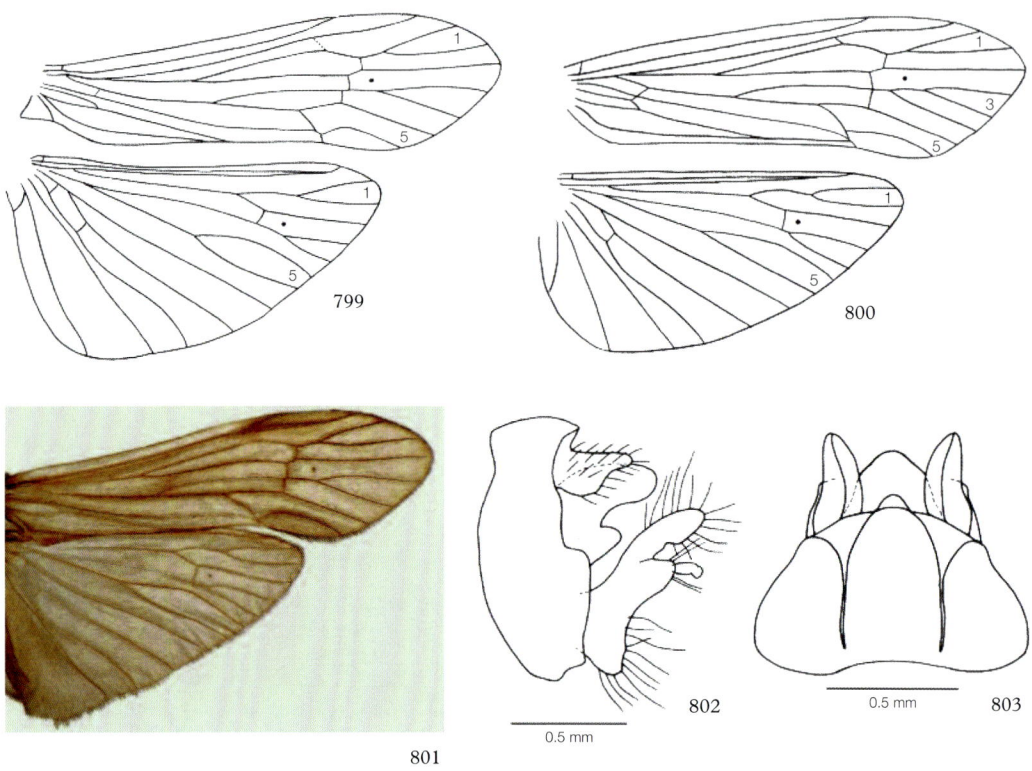

Figures 799-803. *Ceraclea nigronervosa*. 799 male wing venation; 800 female wing venation; 801 wing pattern; 802 male genitalia lateral, aedeagus omitted; 803 female genitalia ventral

## *Ceraclea dissimilis* (Stephens, 1836)

Listed as *Leptocerus dissimilis* in Mosely (1939) and *Arthripsodes* [sic] *dissimilis* in Macan (1973). Fore wing length: ♂ 8-10 mm, ♀ 6-8 mm (Fig. 804); rather uniformly pale brownish, though often darker towards the wing tip. Probably the most common species of *Ceraclea*, found throughout Britain; present in Ireland; rivers and lakes. Throughout Europe. Flight period: June-September. ♂ genitalia as in Fig. 805; ♀ genitalia as in Fig. 806.

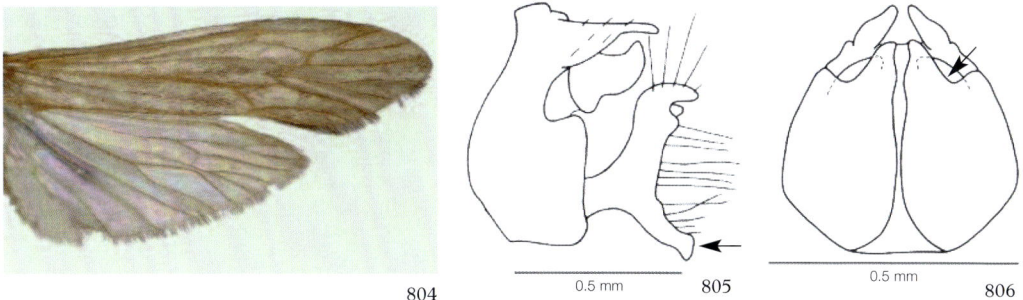

Figures 804-806. *Ceraclea dissimilis*. 804 wing pattern;
805 male genitalia lateral, aedeagus omitted; 806 female genitalia ventral

## *Ceraclea senilis* (Burmeister, 1839)

Listed as *Leptocerus senilis* in Mosely (1939) and *Arthripsodes* [sic] *senilis* in Macan (1973). Fore wing length: ♂♀ 10-14 mm (Fig. 807); uniformly yellowish brown. Common in SE. and E. England as far north as Cheshire and Yorkshire, local in SW. Scotland though it seems to be increasing its range in N. England; present in Ireland; still or slow water. Throughout much of Europe. Flight period: June-September. ♂ genitalia as in Fig. 808; ♀ genitalia as in Fig. 809.

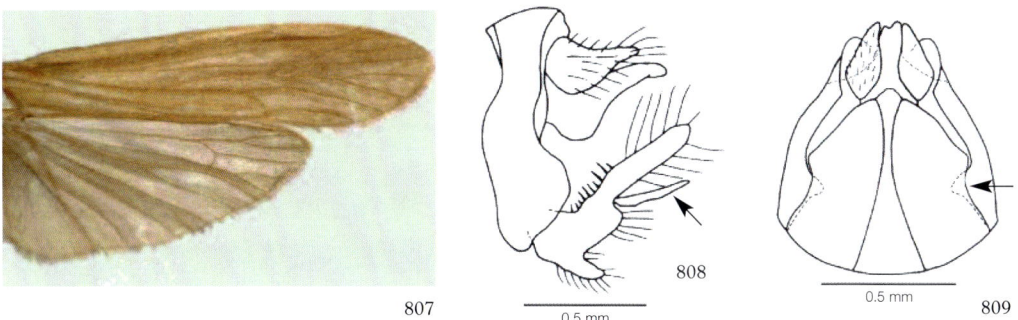

Figures 807-809. *Ceraclea senilis*. 807 wing pattern;
808 male genitalia lateral, aedeagus omitted; 809 female genitalia ventral

# Genus LEPTOCERUS Leach, 1815

Both wings with forks 1 and 5 present; hind wing with a small fold anterior to fork 5 (Fig. 810). Spur formula 0.2.2. About six species in Europe, of which three occur in Britain. Both male and female genitalia give an easy separation of all the British species.

## *Leptocerus tineiformis* Curtis, 1834

Listed as *Setodes tineiformis* in Mosely (1939). Fore wing length: ♂♀ 6-8 mm (Figs 810-812); greyish with small brown spots. Common in S. England and Midlands, more local north and west to its limits in SW. Wales and NW. England; absent from Scotland; present in Ireland; ponds, lakes and canals. Throughout much of Europe. Flight period: June-August. ♂ genitalia as in Fig. 813; ♀ genitalia as in Fig. 814.

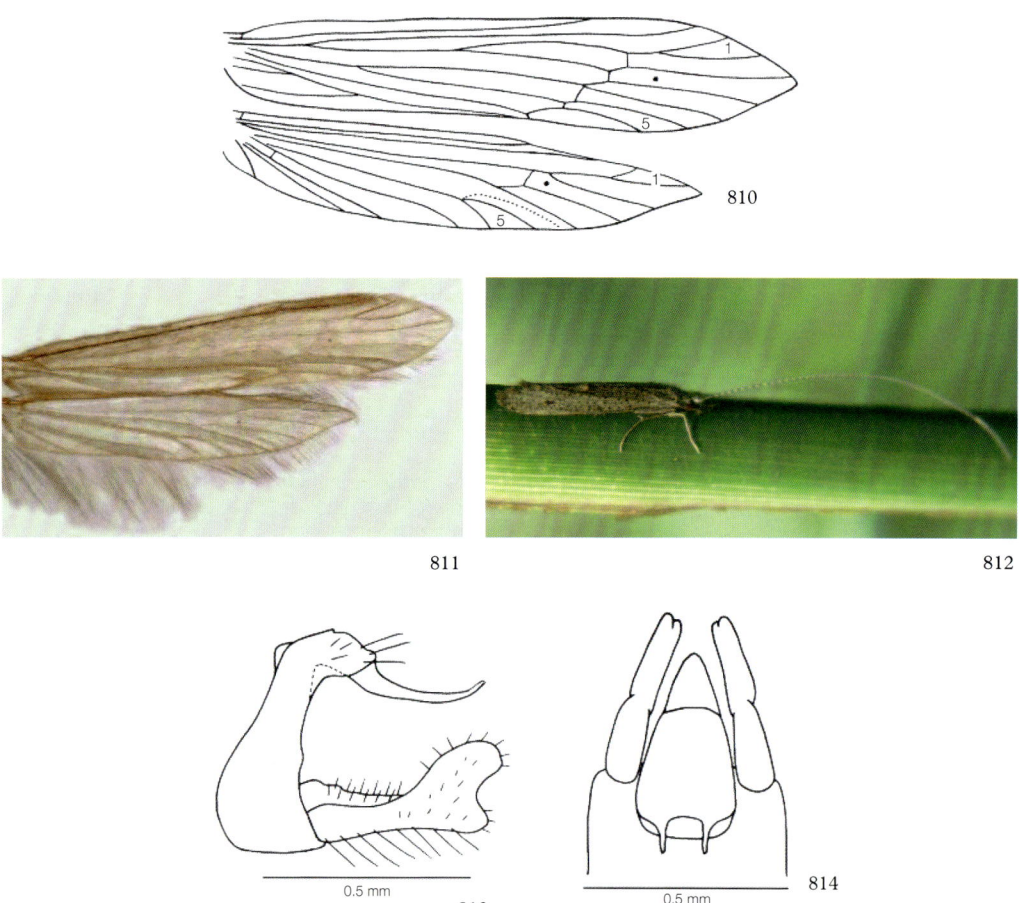

Figures 810-814. *Leptocerus tineiformis*. 810 wing venation; 811 wing pattern; 812 live specimen; 813 male genitalia lateral, aedeagus omitted; 814 female genitalia ventral

## *Leptocerus interruptus* (Fabricius, 1775)

Listed as *Setodes interrupta* in Mosely (1939). Fore wing length: ♂ 6-7 mm, ♀ 5-6 mm (Fig. 815); black with white markings. Very local in SW. England, SE. Wales and adjoining English counties; no records from Ireland; rivers. Mainly southern and central Europe. Flight period: July-August. ♂ genitalia as in Fig. 816; ♀ genitalia as in Fig. 817.

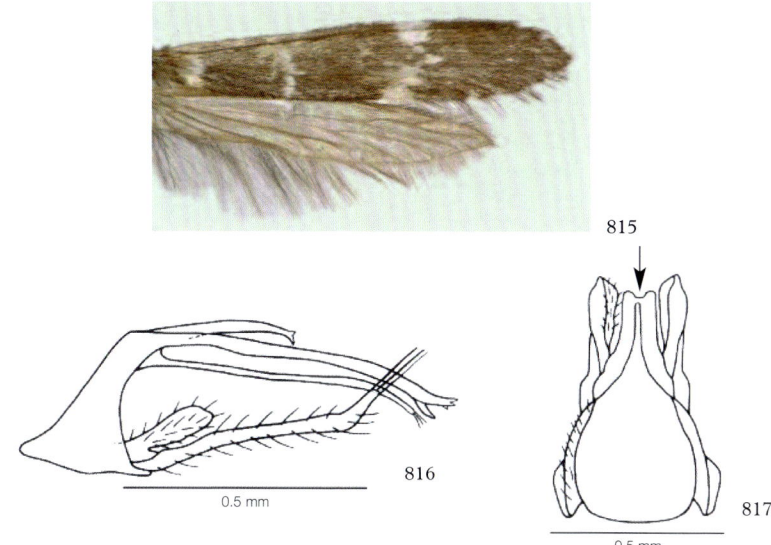

Figures 815-817. *Leptocerus interruptus*. 815 wing pattern;
816 male genitalia lateral, aedeagus omitted; 817 female genitalia ventral

## *Leptocerus lusitanicus* (McLachlan, 1884)

Listed as *Setodes lusitanica* in Mosely (1939). Fore wing length: ♂ ♀ 6-7 mm (Fig. 818); grey with dark spots, similar to *L. tineiformis*. Very local but apparently extending its range: the Dorset Stour and the River Thames with associated gravel pits, and may be becoming established in several East Anglian rivers; no records from Ireland; large rivers, gravel pits. Mainly south-western Europe. Flight period: July-August. ♂ genitalia as in Fig. 819; ♀ genitalia as in Fig. 820.

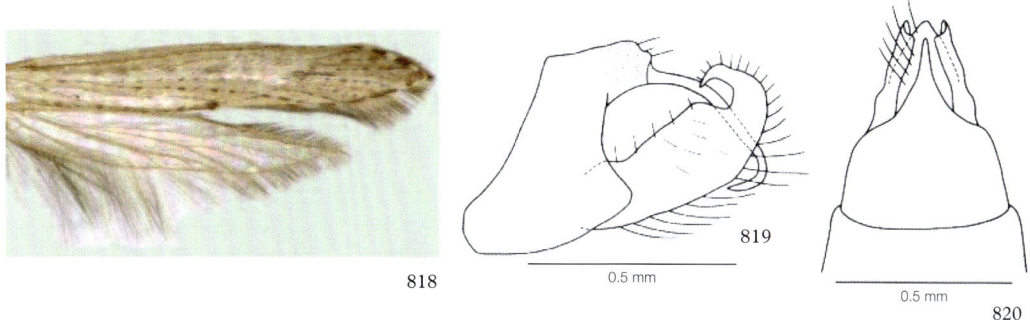

Figures 818-820. *Leptocerus lusitanicus*. 818 wing pattern;
819 male genitalia lateral, aedeagus omitted; 820 female genitalia ventral

# Genus MYSTACIDES Latreille, 1825

Both wings with forks 1 and 5 present (Fig. 825). Spur formula 0.2.2. The three species in Britain are also the only ones in Europe. Characters to separate the females were provided by Nógrádi (1997). All three British species have conspicuously red eyes and 'elbowed' maxillary palps.

## *Mystacides azurea* (Linnaeus, 1761)
= form *albicornis* Mosely, 1930

This species and *M. nigra* are the angler's Black Silverhorns; in both species the tip of each fore wing bends sharply at the line of cross-veins, giving a characteristic angular shape to the outline in dorsal view (Fig. 821). Fore wing length: ♂♀ 6-9 mm (Fig. 821); shiny bluish black in life, fading to brown after death. Common throughout Britain; present in Ireland; streams, rivers, canals, lakes and large ponds. Throughout Europe. Flight period: May-September. ♂ genitalia with Y-shaped ventral plate (Figs 822, 823); ♀ genitalia with subrectangular ventral appendages (Fig. 824).

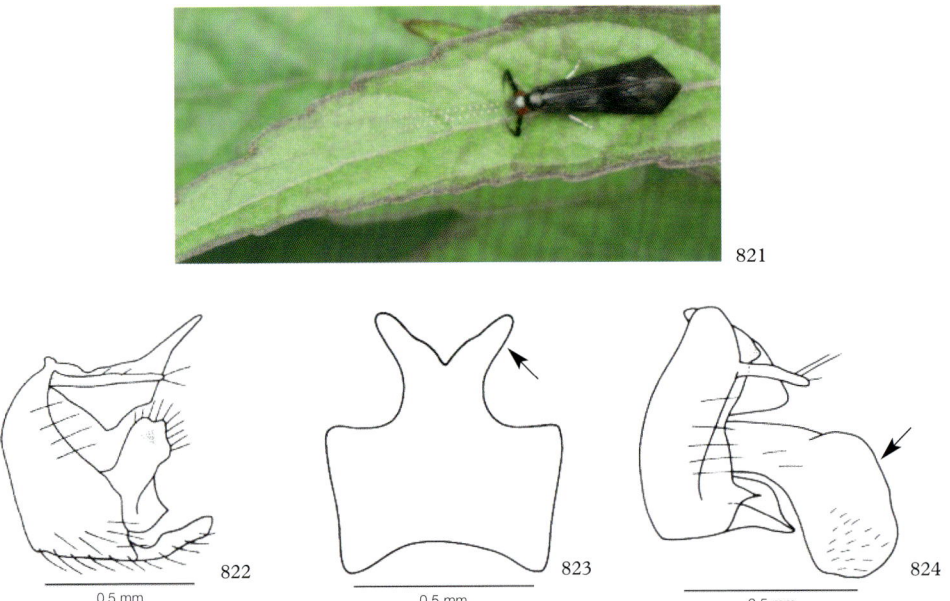

Figures 821-824. *Mystacides azurea*. 821 live specimen;
822 male genitalia lateral, aedeagus omitted; 823 male genitalia ventral; 824 female genitalia lateral

## *Mystacides nigra* (Linnaeus, 1758)

This species and *M. azurea* are the angler's Black Silverhorns; in both species the tip of each fore wing bends sharply at the line of cross-veins, giving a characteristic angular shape to the outline in dorsal view (Fig. 827). Fore wing length: ♂♀ 7-10 mm (Figs 825-827); black in life, fading to brown after death; usually less shiny in appearance than *M. azurea*. Local throughout England, Wales and mainland Scotland; no records from Ireland; streams, rivers, canals and lakes. Throughout much of Europe except south. Flight period: May-September. ♂ genitalia with elongate ventral plate that has slight apical excision (Figs 828, 829); ♀ genitalia with rounded excision in posterior margin of ventral appendage (Fig. 830).

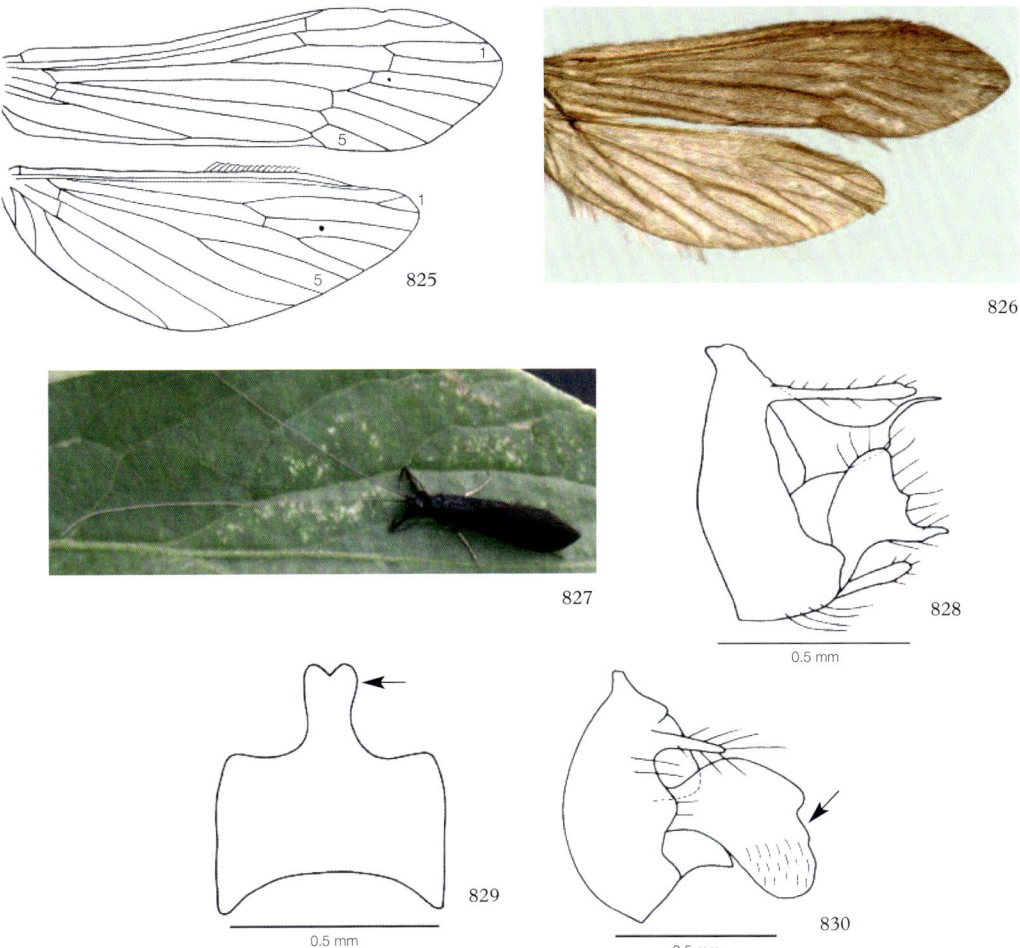

Figures 825-830. *Mystacides nigra*. 825 wing venation; 826 wing pattern; 827 live specimen; 828 male genitalia lateral, aedeagus omitted; 829 male genitalia ventral; 830 female genitalia lateral

## *Mystacides longicornis* (Linnaeus, 1758)

This is the angler's Grouse Wing. Fore wing length: ♂ ♀ 6-9 mm (Figs 831-833). Two colour varieties are known: as well as the typical striking brown and yellow striped form (Fig. 832), the pale yellow form is moderately common (Fig. 833). Throughout Britain; present in Ireland; ponds, lakes, slow rivers and canals. Throughout much of Europe except south. Flight period: May-September (October). ♂ genitalia with elongate ventral process that has simple rounded apex (Figs 834, 835); ♀ genitalia with enlarged mid-dorsal section of ventral appendage (Fig. 836).

Figures 831-836. *Mystacides longicornis*. 831 wing pattern; 832 typical live specimen; 833 pale form; 834 male genitalia lateral, aedeagus omitted; 835 male genitalia ventral; 836 female genitalia lateral

# Genus SETODES Rambur, 1842

Both wings with forks 1 and 5 present (Fig. 837). Spur formula 0.2.2. Of about 15 European species only two are found in Britain. These small species are easily overlooked, though they are attracted to light.

## *Setodes argentipunctellus* McLachlan, 1877

Listed as *Setodes argentipunctella* in Mosely (1939). Fore wing length: ♂ 4-6 mm, ♀ 5-6 mm (Figs 837, 838); yellowish grey with silvery white spots. Very local in rivers in SW. England, a few lakes in the Lake District and SW. Scotland; present in Ireland; stony lakes and rivers. Throughout much of western Europe. Flight period: (June) July-August (September). ♂ genitalia as in Fig. 839; ♀ genitalia as in Fig. 840.

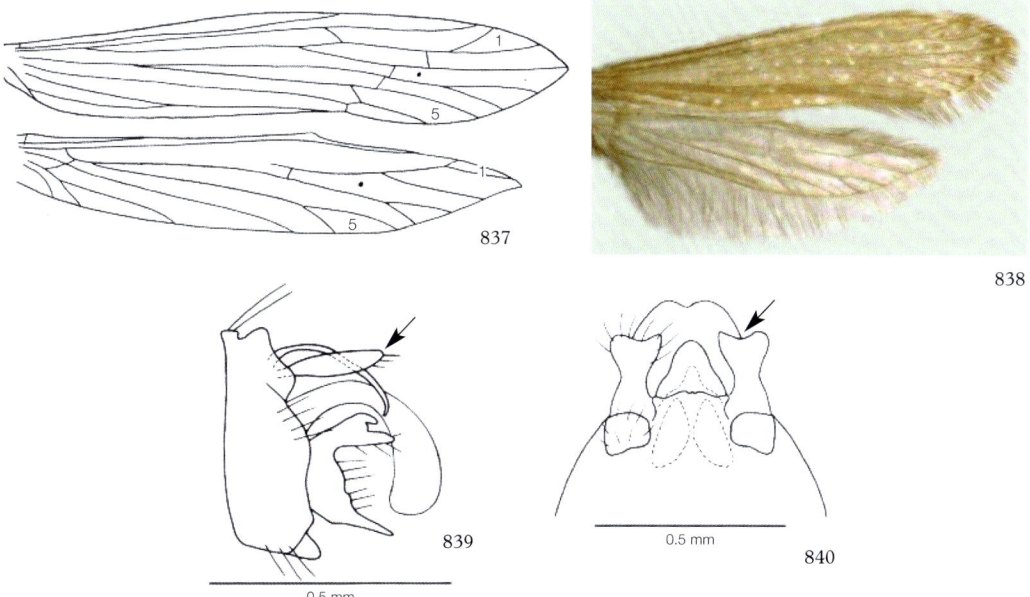

Figures 837-840. *Setodes argentipunctellus*. 837 wing venation; 838 wing pattern;
839 male genitalia lateral; 840 female genitalia ventral

## *Setodes punctatus* (Fabricius, 1793)

Listed as *Setodes punctata* in Mosely (1939). Fore wing length: ♂ 6-7 mm, ♀ 5-7 mm (Fig. 841); similar to *S. argentipunctellus* though white spots often larger. Rivers Wye and Severn only; no records from Ireland; large rivers. Throughout much of Europe, less common in the north. Flight period: July-September. ♂ genitalia as in Fig. 842; ♀ genitalia as in Fig. 843.

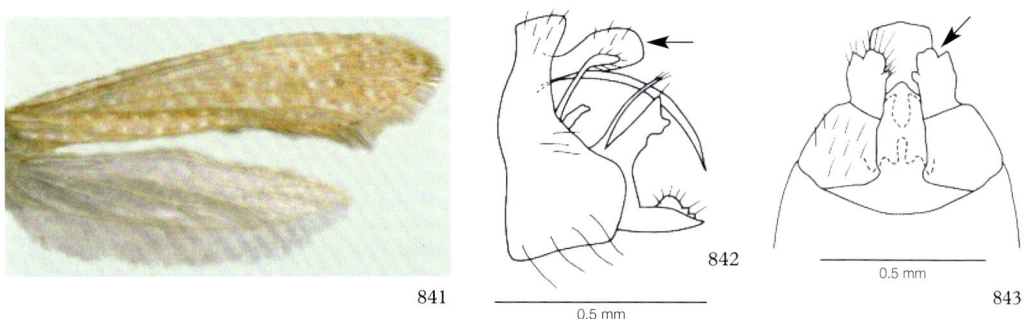

Figures 841-843. *Setodes punctatus*. 841 wing pattern;
842 male genitalia lateral, aedeagus omitted; 843 female genitalia ventral

# Genus OECETIS McLachlan, 1877

Both wings with forks 1 and 5 present (Fig. 844). Spur formula 0.2.2 or 1.2.2. Of over 12 species in Europe five are found in Britain. These are generally known to anglers as the Longhorn Sedges.

## *Oecetis ochracea* (Curtis, 1825)

Fore wing length: ♂ 10-13 mm, ♀ 9-11 mm (Figs 844-846); a large and very pale species. Spur formula 1.2.2. Common throughout Britain; present in Ireland; ponds, lakes, canals, slow rivers. Throughout Europe. Flight period: May-September; often found in light-traps at some distance from water-bodies and it is probably a coloniser of new large ponds and lakes. ♂ genitalia as in Fig. 847; ♀ genitalia as in Fig. 848.

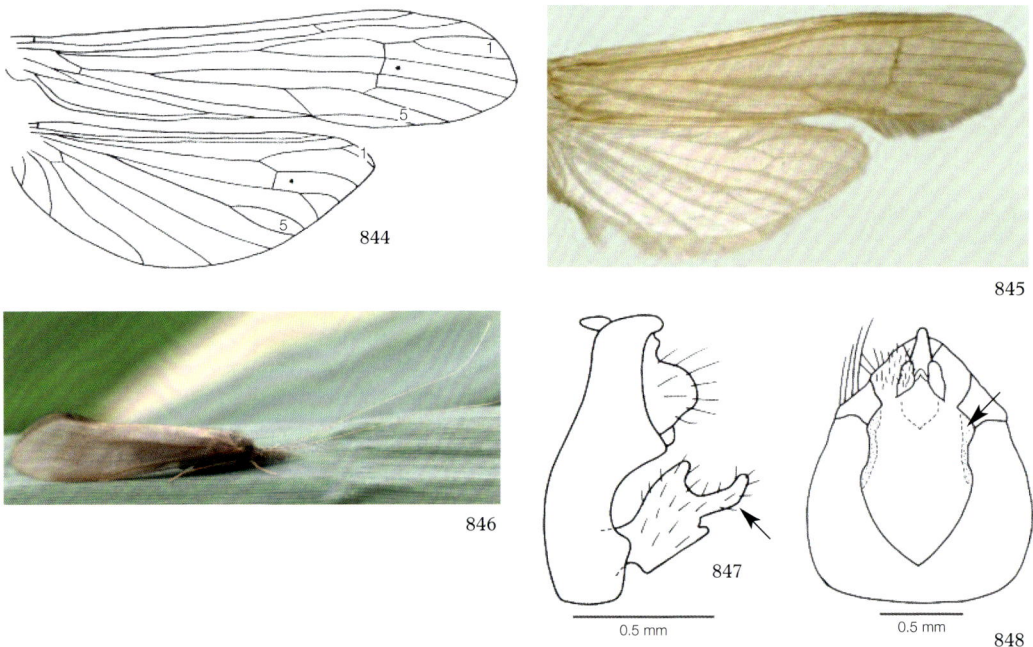

Figures 844-848. *Oecetis ochracea*. 844 wing venation; 845 wing pattern; 846 live specimen;
847 male genitalia lateral, aedeagus omitted; 848 female genitalia ventral

## *Oecetis lacustris* (Pictet, 1834)

Fore wing length: ♂ 6-7 mm, ♀ 6-8 mm (Fig. 849); greyish yellow with some dark spots. Spur formula 0.2.2. Common throughout most of Britain, but absent from N. Scotland; present in Ireland; ponds lakes, canals, slow rivers. Throughout most of Europe except south. Flight period: June-September. ♂ genitalia as in Fig. 850; ♀ genitalia as in Fig. 851.

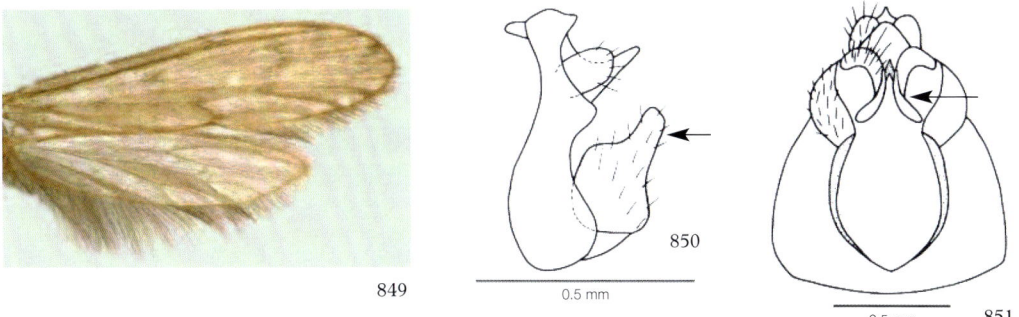

Figures 849-851. *Oecetis lacustris*. 849 wing pattern;
850 male genitalia lateral, aedeagus omitted; 851 female genitalia ventral

## *Oecetis furva* (Rambur, 1842)

Fore wing length: ♂ 7-8 mm, ♀ 8-10 mm (Fig. 852); reddish, sometimes resembling *Triaenodes bicolor*, Fig. 756. Spur formula 0.2.2. Throughout Britain, but very few records; present in Ireland; large ponds and lakes, ditches; always in habitats with much marginal and submerged vegetation. Throughout most of Europe except south. Flight period: (May) June-September. ♂ genitalia as in Fig. 853; ♀ genitalia as in Fig. 854.

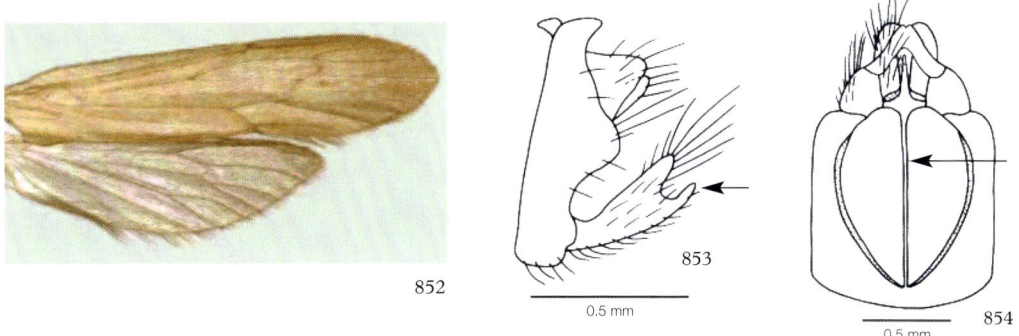

Figures 852-854. *Oecetis furva*. 852 wing pattern;
853 male genitalia lateral, aedeagus omitted; 854 female genitalia ventral

## *Oecetis notata* (Rambur, 1842)

Fore wing length: ♂ 7-8 mm, ♀ 7-9 mm (Figs 855, 856); although the pattern of dark lines and spots is not very distinctive it is diagnostic for this species. Spur formula 1.2.2. Local in Wales and bordering English counties, with older records from the Thames and tributaries; present in Ireland; fast rivers. Throughout much of Europe. Flight period: June-August. ♂ genitalia as in Fig. 857; ♀ genitalia as in Fig. 858.

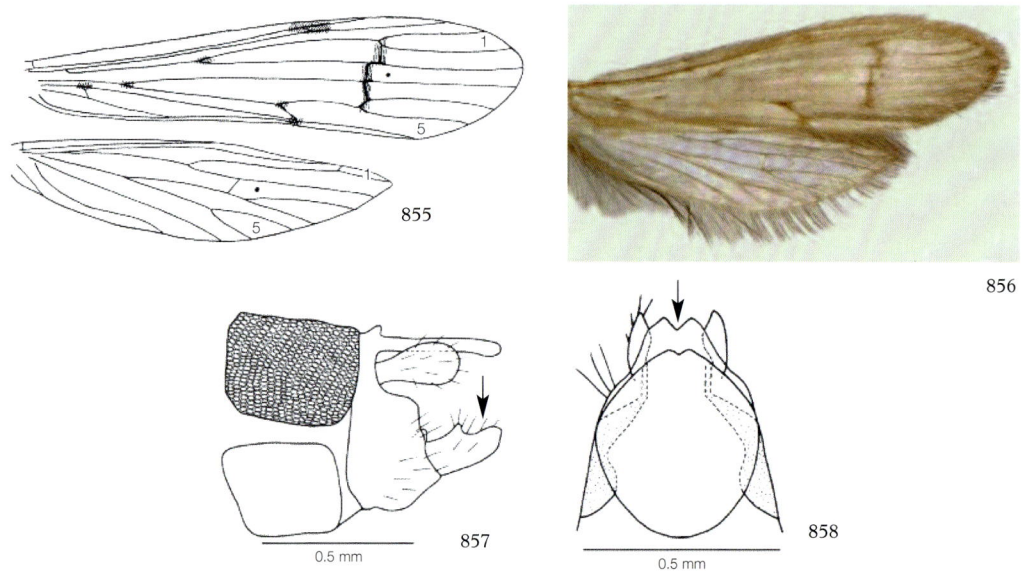

Figures 855-858. *Oecetis notata*. 855 wing venation; 856 wing pattern;
857 male genitalia lateral, aedeagus omitted; 858 female genitalia ventral

## *Oecetis testacea* (Curtis, 1834)

Fore wing length: ♂♀ 6-9 mm (Fig. 859); yellow with indistinct darker markings. Spur formula 1.2.2. Throughout Britain, but apparently absent from E. England; present in Ireland; streams, rivers, canals and stony lakes. Throughout much of Europe. Flight period: June-August. ♂ genitalia as in Fig. 860; ♀ genitalia as in Fig. 861.

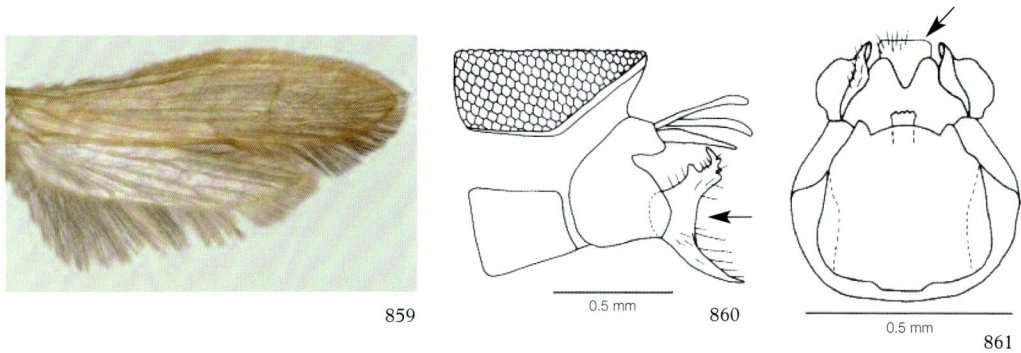

Figures 859-861. *Oecetis testacea*. 859 wing pattern;
860 male genitalia lateral, aedeagus omitted; 861 female genitalia ventral

# References

Ashe, P., O'Connor, J.P. & Murray, D.A.1998. Order Trichoptera. In: A checklist of Irish aquatic insects. *Occasional Publications of the Irish Biogeographical Society* **3**: 27-32.

Badcock, R.M. 1977. The *Hydropsyche fulvipes-instabilis-saxonica* (Trichoptera) complex in Britain and the recognition of *H. siltalai* Döhler. *Entomologist's Monthly Magazine* **113**: 23-29.

Badcock, R.M. 1978. Taxonomic controversies in the Hydropsychidae. Pp 175-182. In: Crichton. M.I. (ed.) *Proceedings of the 2nd International Symposium on Trichoptera*. Junk, The Hague.

Barnard, P.C. 1978. Oviposition in the caddis-fly *Brachycentrus subnubilus* Curtis (Trichoptera, Brachycentridae). *Entomologist's Gazette* **29**: 159-161.

Barnard, P.C. 1985. An annotated check-list of the Trichoptera of Britain and Ireland. *Entomologist's Gazette* **36**: 31-45.

Barnard, P.C. & O'Connor, J.P. 1987. The populations of *Apatania muliebris* McLachlan in the British Isles (Trichoptera: Limnephilidae). *Entomologist's Gazette* **38**: 263-268.

Barnard, P. & Ross, E. 2008. *Guide to the adult caddisflies or sedge flies (Trichoptera)*. Field Studies Council, OP 129.

Cooling, D.A. 1982. Records of Trichoptera from rivers in southern England. *Entomologist's Gazette* **33**: 123-134.

Crichton, M.I. 1971. A study of caddis flies (Trichoptera) of the family Limnephilidae, based on the Rothamsted Insect Survey, 1964-68. *Journal of Zoology, London* **163**: 533-563.

Crichton, M.I. 1987. A study of egg masses of *Glyphotaelius pellucidus* (Retzius) (Trichoptera: Limnephilidae). In: Bournaud, M. & Tachet, H. (eds) *Proceedings of the 5th International Symposium on Trichoptera*. Dordrecht: pp 165-169.

Crichton, M.I. 1991. A scanning electron microscope study of the mouth parts of adult *Phryganea grandis* (L.) (Trichoptera). In: Tomaszewski, C. (ed.) *Proceedings of the 6th International Symposium on Trichoptera*. Poznan, Poland: pp 329-333.

Crichton, M.I. & Fisher, D.B. 1981. Further observations on limnephilid life histories, based on the Rothamsted Insect Survey. In: Moretti, G.P. (ed.) *Proceedings of the 3rd International Symposium on Trichoptera*. The Hague: pp 47-55.

Crichton, M.I. & Fisher, D.B. 1982. Records of caddis flies (Trichoptera) from Rothamsted light traps at Field Centres. *Field Studies* **5**: 569-579.

Crichton, M.I., Fisher, D.B. & Woiwod, I.P. 1978. Life histories and distribution of British Trichoptera, excluding Limnephilidae and Hydroptilidae, based on the Rothamsted Insect Survey. *Holarctic Ecology* **1**: 31-45.

Crofts, S.M. 2011. *Synagapetus dubitans* McLachlan, 1879 (Trichoptera, Glossosomatidae), a caddisfly new to Great Britain. *Entomologist's Monthly Magazine* **147**: 32.

Décamps, H. 1966. Sur la détermination des femelles de *Potamophylax latipennis* (Curt.) Neboiss et *Potamophylax cingulatus* (Stephens) (Trichoptera). *Annales de Limnologie* **2**: 537-541.

Denis, C. 1981. Action de la photopériode sur la maturation génitale des femelles de quelques Limnéphilidés. In: Moretti, G.P. (ed.) *Proceedings of the 3rd International Symposium on Trichoptera*. The Hague: pp 57-66.

Drake, C.M. & Willo, A. 2009. *Hydropsyche bulgaromanorum* Malicky (Trichoptera: Hydropsychidae) rediscovered in England. *British Journal of Entomology and Natural History* **22**: 23-29.

Dreesmann, D.C. & Wichard, W. 2002. The basal phylogenetical relationships of Trichoptera – a molecular approach. In: Mey, W. (ed.) *Proceedings of the 10th International Symposium on Trichoptera*. Keltern, Germany: pp 309-316.

Edington, J.M. & Hildrew, A.G. 1995. A revised key to the caseless caddis larvae of the British Isles with notes on their ecology. *Scientific Publications of the Freshwater Biological Association* **53**: 134 pp.

Fisher, D. 1977. Identification of adult females of *Tinodes* in Britain (Trichoptera: Psychomyiidae). *Systematic Entomology* **2**: 105-110.

Fox, M.W. 1957. Abnormalities in neuration in Trichoptera. *Entomologist's Monthly Magazine* **93**: 237-238.

Gower, A.M. 1965. The life cycle of *Drusus annulatus* Steph. (Trich., Limnephilidae) in watercress beds. *Entomologist's Monthly Magazine* **101**: 133-141.

Green, H. & Westwood, B. 2005. In search of the land caddis. *British Wildlife* **17**: 21-26.

Grensted, L.W. 1939. *Ithytrichia clavata* Morton (Trich., Hydroptilidae) new to Britain. *Entomologist's Monthly Magazine* **74**: 235.

Grensted, L.W. 1943. The occurrence of *Hydropsyche saxonica* McLach. in Britain, with a new key to the British species of the genus *Hydropsyche* Pict. (Trich., Hydropsychidae). *Entomologist's Monthly Magazine* **79**: 35-38.

Grigorenko, V.N. 2002. Some taxonomical notes on the limnephiline caddisflies (Trichoptera: Limnephilidae, Limnephilinae). In: Mey, W. (ed.) *Proceedings of the 10th International Symposium on Trichoptera*. Keltern, Germany: pp 107-119.

Gullefors, B. 2005. Trichoptera from the brackish water of the Gulf of Bothnia. In: Tanida, K. & Rossiter, A. (eds) *Proceedings of the 11th International Symposium on Trichoptera*. Kanagawa, Japan: pp 137-147.

Hickin, N.E. 1967. *Caddis larvae: larvae of the British Trichoptera*. Hutchinson, London. 476 pp.

Hildrew, A.G. & Morgan, J.C. 1974. The taxonomy of the British Hydropsychidae. *Journal of Entomology* (B) **43**: 217-229.

Hoffmann, A. 1999. Mating systems in Trichoptera: a little about the little known. In: Malicky, H. & Chantaramongkol, P. (eds) *Proceedings of the 9th International Symposium on Trichoptera*. Chiang Mai, Thailand: pp 133-139.

Holzenthal, R.W., Blahnik, R.J., Kjer, K.M. & Prather, A.L. 2007. An update on the phylogeny of caddisflies (Trichoptera). In: Bueno-Soria, J., Barba-Álvarez, R. & Armitage, B.J. (eds) *Proceedings of the 12th International Symposium on Trichoptera*. The Caddis Press, Columbus, Ohio: pp 143-153.

Invertebrate Link (JCCBI) (2002) A code of conduct for collecting insects and other invertebrates. *British Journal of Entomology and Natural History* **15**: 1-6.

Ivanov, V.D. 1997. Vibrations, pheromones, and communication patterns in Trichoptera. In: Holzenthal, R.W. & Flint, O.S. Jr (eds) *Proceedings of the 8th International Symposium on Trichoptera*. Columbus, Ohio: pp 183-188.

Ivanov, V.D. 2005. Ground plan and basic evolutionary trends of male terminal segments in Trichoptera. In: Tanida, K. & Rossiter, A. (eds) *Proceedings of the 11th International Symposium on Trichoptera*. Kanagawa, Japan: pp 207-218.

Ivanov, V.D. & Löfstedt, C. 1999. Pheromones in caddisflies. In: Malicky, H. & Chantaramongkol, P. (eds) *Proceedings of the 9th International Symposium on Trichoptera*. Chiang Mai, Thailand: pp 149-156.

Ivanov, V.D. & Melnitsky, S.I. 2002. Structure of pheromone glands in Trichoptera. Pp 17-28. In: Mey, W. (ed.) *Proceedings of the 10th International Symposium on Trichoptera*. Keltern, Germany.

Kimmins, D.E. 1942. *Cyrnus insolutus* McL. (Trichoptera), new to Britain. *Entomologist* **75**: 66-68.

Kimmins, D.E. 1949. *Tinodes pallidula* McLachlan, an addition to the British list of Trichoptera. *Entomologist* **82**: 269-272.

Kimmins, D.E. 1952. *Agrypnetes crassicornis* McLachlan (Fam. Phryganeidae), a caddisfly new to Britain. *Annals and Magazine of Natural History* (12) **5**: 1039-1043.

Kimmins, D.E. 1953. A key to the European species of *Wormaldia* (Trichoptera, Philopotamidae) with descriptions of two new subspecies. *Annals and Magazine of Natural History* (12) **6**: 801-808.

Kimmins, D.E. 1956. British Trichoptera (caddis flies). *Entomologist's Gazette* 7: 29-38.

Kimmins, D.E. 1961. A species of *Hydroptila* (Trichoptera) new to Britain. *Entomologist's Gazette* **12**: 32-35.

Kimmins, D.E. 1964. *Triaenodes simulans* Tjeder in Britain (Trichoptera, Leptoceridae). *Entomologist* **97**: 40-44.

Kimmins, D.E. 1966. A revised checklist of the British Trichoptera. *Entomologist's Gazette* **17**: 111-120.

Macan, T.T. 1973. A key to the adults of the British Trichoptera. *Scientific Publications of the Freshwater Biological Association* **28**: 151 pp.

Malicky, H. 1984a. The Atlas of European Trichoptera, and some thoughts about identification of specimens. In: Morse, J.C. (ed.) *Proceedings of the 4th International Symposium on Trichoptera*. The Hague: pp 203-205.

Malicky, H. 1984b. The distribution of *Hydropsyche guttata* Pictet and *H. bulgaromanorum* Malicky (Trichoptera: Hydropsychidae), with notes on their bionomics. *Entomologist's Gazette* **35**: 257-264.

Malicky, H. 1991. Life cycle strategies in some European caddisflies. In: Tomaszewski, C. (ed.) *Proceedings of the 6th International Symposium on Trichoptera*. Poznan, Poland: pp 195-197.

Malicky, H. 1995. Eine neue *Psychomyia* aus dem südöstlichen Mitteleuropa, mit Bemerkungen über die Gattung *Metalype* (Trichoptera: Psychomyiidae). *Entomologische Zeitschrift* **105**: 441-446.

Malicky, H. 1998. Revision der Gattung *Mesophylax* McLachlan (Trichoptera, Limnephilidae). *Beiträge zur Entomologie* **48**: 115-144.

Malicky, H. 2002. The sub-specific division of *Rhyacophila dorsalis* Curtis, 1834 and its transitions to *R. nubila* Zetterstedt, 1840 (Trichoptera: Rhyacophilidae). In: Mey, W. (ed.) *Proceedings of the 10th International Symposium on Trichoptera*. Keltern, Germany: pp 149-166.

Malicky, H. 2004. *Atlas of European Trichoptera*. 2nd edition. Springer, Dordrecht.

Malicky, H. 2005. Ein kommentiertes Verzeichnis der Köcherfliegen (Trichoptera) Europas und des Mediterrangebietes. *Linzer Biologische Beiträge* **37**: 533-596.

Marshall, J.E. 1977. *Hydroptila martini* sp. n. and *Hydroptila valesiaca* Schmid (Trichoptera: Hydroptilidae) new to the British Isles. *Entomologist's Gazette* **28**: 115-122.

Marshall, J.E. 1978. Trichoptera: Hydroptilidae. *Handbooks for the Identification of British Insects* **1** (14a): iv + 31 pp.

Marshall, J.E. 1979. A description of the female of *Hydroptila tigurina* Ris (Trichoptera: Hydroptilidae). *Entomologist's Gazette* **30**: 213-214.

McLachlan, R. 1874-1880. *A monographic revision and synopsis of the Trichoptera of the European fauna*. London.

Mey, W. 2007. The value of the female genitalia for the systematics of the genus *Hydropsyche* Pictet, 1834 (Trichoptera: Hydropsychidae). In: Bueno-Soria, J., Barba-Álvarez, R. & Armitage, B.J. (eds) *Proceedings of the 12th International Symposium on Trichoptera*. The Caddis Press, Columbus, Ohio: pp 203-210.

Morse, J.C. & Chuluunbat, S. 2007. Skating caddisflies of Mongolia. In: Bueno-Soria, J., Barba-Álvarez, R. & Armitage, B.J. (eds) *Proceedings of the 12th International Symposium on Trichoptera*. The Caddis Press, Columbus, Ohio: pp 219-227.

Morse, J.C. & Wallace, I.D. 1976. *Athripsodes* Billberg and *Ceraclea* Stephens, distinct genera of long-horned caddis-flies (Trichoptera: Leptoceridae). In: Malicky, H. (ed.) *Proceedings of the First International Symposium on Trichoptera*, pp. 33-40. Junk, The Hague.

Morton, K.J. 1906. *Triaenodes reuteri* McLach., a species of Trichoptera new to Britain. *Entomologist's Monthly Magazine* **42**: 270-271.

Morton, K.J. 1931. *Triaenodes simulans* Tjeder, a species of Trichoptera new to Britain, and a correction. *Entomologist's Monthly Magazine* **67**: 16-17.

Mosely, M.E. 1921. *The dry-fly fisherman's entomology*. Routledge, London.

Mosely, M.E. 1939. *The British caddis flies (Trichoptera)*. Routledge, London.

Moss, M.O. & Gibbs, G. 2000. On the nature of the hairs of the wings of the Trichoptera (caddisflies). *Quekett Journal of Microscopy* **38**: 511-517.

Neboiss, A. 1963. The Trichoptera types of species described by J. Curtis. *Beiträge zur Entomologie* **13**: 582-635.

Neu, P.J. & Tobias, W. 2004. Die Bestimmung der in Deutschland vorkommenden Hydropsychidae (Insecta: Trichoptera). *Lauterbornia* **51**: 1-68.

Nógrádi, S. 1997. How to distinguish the females of the three European *Mystacides* species. *Braueria* **24**: 18.

O'Connor, J.P. 1978. *Hydroptila tigurina* Ris new to Ireland with notes on *Apatania wallengreni* McLachlan and *Limnephilus binotatus* Curtis (Insecta: Trichoptera). *Irish Naturalists' Journal* **19**: 191-192.

O'Connor, J.P. 1980. *Limnephilus pati* sp.n. (Trichoptera: Limnephilidae) a caddis fly new to Great Britain and Ireland. *Irish Naturalists' Journal* **20**: 129-133.

O'Connor, J.P. 2010. Additions and corrections to the Irish list of Trichoptera. *Irish Naturalists' Journal* **30**: 127-129.

O'Connor, J.P. & Barnard, P.C. 1981. *Limnephilus tauricus* Schmid (Trichoptera: Limnephilidae) new to Great Britain, with a key to the *L. hirsutus* (Pictet) group in the British Isles. *Entomologist's Gazette* **32**: 115-119.

O'Connor, J.P. & Wise, E.J. 1980. The larva of *Tinodes maculicornis* (Trichoptera: Psychomyiidae), with notes on the species' distribution in Ireland. *Freshwater Biology* **10**: 367-370.

Oláh, J. & Johanson, K.A. 2007. Trinominal terminology for cephalic setose warts in Trichoptera (Insecta). *Braueria* **34**: 43-50.

Pelham-Clinton, E.C. 1966a. *Nemotaulius punctatolineatus* (Retzius), a caddis-fly new to the British Isles (Trichoptera, Limnephilidae). *Entomologist's Gazette* **17**: 5-8.

Pelham-Clinton, E.C. 1966b. *Triaenodes reuteri* McLachlan, a species to be restored to the British list; and a redescription of *T. simulans* Tjeder (Trichoptera, Leptoceridae). *Entomologist* **99**: 47-50.

Petersson, E. 1991. Polyandry in some caddis flies (Trichoptera). In: Tomaszewski, C. (ed.) *Proceedings of the 6th International Symposium on Trichoptera*. Poznan, Poland: pp 213-215.

Rathjen, W. 1939. Experimentelle Untersuchungen zur Biologie und Ökologie von *Enoicyla pusilla* Burm. *Zeitschrift für Morphologie und Ökologie der Tiere* **35**: 14-84.

Robert, B. & Neu, P.J. 2002. Characters for distinguishing *Cyrnus*-females (Trichoptera: Polycentropodidae) in northern, eastern and most parts of central Europe. In: Mey, W. (ed.) *Proceedings of the 10th International Symposium on Trichoptera*. Keltern, Germany: pp 235-238.

Robert, B. & Schmidt, C. 1990. Zur Unterscheidung der Weibchen von *Potamophylax cingulatus* (Stephens 1837) und *Potamophylax latipennis* (Curtis 1834) (Trichoptera: Limnephilidae). *Entomologische Zeitschrift* **100**: 293-312.

Rojas-Camousseight, F. & Tachet, H. 1988. Les femelles d'*Hydroptila* du groupe *sparsa* (Trichoptera, Hydroptilidae). *Rivista di Idrobiologia* **27**: 309-316.

Rojas-Camousseight, F., Usseglio-Polatera, P., Tachet, H. & Bournaud, M. 1991. The identification of females of *Hydropsyche* (Trichoptera, Hydropsychidae): a puzzle for the ecologist? In: Tomaszewski, C. (ed.) *Proceedings of the 6th International Symposium on Trichoptera*. Poznan, Poland: pp 323-327.

Ross, E. 2006. Caddis flies (Trichoptera) collected from Wicken Fen, Cambridgeshire and Minsmere, Suffolk during 2004-2005. *British Journal of Entomology and Natural History* **19**: 206-208.

Ross, E. 2008. Caddisflies (Trichoptera) collected in the Malham Tarn area (Yorkshire) in July 2006 and 2007. *British Journal of Entomology and Natural History* **21**: 152-154.

Ross, H.H. 1944. The caddis flies, or Trichoptera, of Illinois. *Bulletin of the Illinois Natural History Survey* **23**: 326 pp.

Roy, D., Décamps, H. & Harper, P.P. 1980. Taxonomy of male and female *Plectrocnemia* (Trichoptera: Polycentropodidae) from the French Pyrenees. *Aquatic Insects* **2**: 19-31.

Schmid, F. 1998. *The insects and arachnids of Canada. Part 7. Genera of the Trichoptera of Canada and adjoining or adjacent United States.* NRC Press, Ottawa. 319pp.

Solem, J.O. & Bongard, T. 1987. Flight patterns of three species of lotic caddisflies. In: Bournaud, M. & Tachet, H. (eds) *Proceedings of the 5th International Symposium on Trichoptera.* Dordrecht: pp 223-228.

Syrnikov, Y.S., Melnitsky, S.I & Ivanov, V.D. 2005. Effect of feeding on the mating behaviour of Trichoptera. Pp 413-419. In: Tanida, K. & Rossiter, A. (eds) *Proceedings of the 11th International Symposium on Trichoptera.* Kanagawa, Japan.

Tindall, A.R. 1963. Keys for the identification of adults of the genus *Limnephilus* (Trich., Limnephilidae). *Entomologist's Monthly Magazine* **99**: 115-123.

Vshivkova, T.S., Morse, J.C. & Ruiter, D. 2007. Phylogeny of Limnephilidae and composition of the genus *Limnephilus* (Limnephilidae: Limnephilinae, Limnephilini). In: Bueno-Soria, J., Barba-Álvarez, R. & Armitage, B.J. (eds) *Proceedings of the 12th International Symposium on Trichoptera.* The Caddis Press, Columbus, Ohio: pp 309-319.

Wallace, I.D. 1991. A review of the Trichoptera of Great Britain. *Research and Survey in Nature Conservation* **32**: 59 pp.

Wallace, I.D., Wallace, B. & Philipson, G.N. 2003 Keys to the case-bearing caddis larvae of Britain and Ireland. *Scientific Publications of the Freshwater Biological Association* **61**: 259pp.

Weaver, J. S. 2002. A synonymy of the caddisfly genus *Lepidostoma* Rambur (Trichoptera: Lepidostomatidae), including a species checklist. *Tijdschrift voor Entomologie* **145**: 173-192.

Whitehead, P.F. 2007. New autecological data for *Enoicyla pusilla* (Burmeister, 1839) (Trichoptera: Limnephilidae) from the Worcestershire Malvern Hills. *Entomologist's Gazette* **58**: 26-28.

Wiggins, G.B. 1998. *The caddisfly family Phryganeidae (Trichoptera).* University of Toronto Press: 306 pp.

# Index

Main entries and start of main sections are shown in **bold**. Synonyms are given in *italics*.

ADICELLA 21, 159, **160**

affinis, Limnephilus 20, **124**, 125

AGAPETUS 15, 30, **33**, 35

AGRAYLEA 16, 36, **45**

AGRYPNETES 18, 91, **92**

AGRYPNIA 18, 91, 93, **94**

albicans, Molanna 21, **157**, 158

albicorne, Odontocerum 4, 21, **156**

*albicornis, Mystacides* 174

albifrons, Athripsodes 22, **164**, 165, 166

*albifrons, Leptocerus* 164

albimacula, Ceraclea 22, **168**, 169

*alboguttatus, Ceraclea* 22, **168**

*alboguttatus, Leptocerus* 168

ALLOGAMUS 20, 119, **146**

ALLOTRICHIA 16, 36, **46**

alpestris, Rhadicoleptus 20, **124**

*alpestris, Stenophylax* 124

ANABOLIA 19, 118, **119**

angulata, Hydroptila 16, **42**

angustata, Molanna 21, **158**

angustella, Orthotrichia 16, **47**, 48

angustipennis, Hydropsyche 18, **90**

annulatus, Drusus 19, **114**

annulicornis, Ceraclea 22, **169**

*annulicornis, Leptocerus* 169

Annulipalpia 14, 17

APATANIA 19, **110**

Apataniidae 14, 19, 24, **110**

*APATIDEA* 110, 111

argentipunctellus, Setodes 22, **177**

articularis, Ernodes 21, **155**

aspersus, Mesophylax 21, **141**

assimilis, Tinodes 18, **80**

*ASYNARCHUS* 139

aterrimus, Athripsodes 22, **167**

*aterrimus, Leptocerus* 167

ATHRIPSODES 22, 159, **164**, 168

*atomarius, Grammotaulius* 20, 122

*aurata, Holocentropus* 64

*aureola, Tinodes* 80

auricollis, Allogamus 20, **146**

*auricollis, Halesus* 146

auricula, Limnephilus 20, **138**

azurea, Mystacides 22, **174**, 175

baltica, Erotesis 22, **160**

basale, Lepidostoma 19, **109**

*basalis, Lasiocephala* 109

BERAEA 21, **152**

Beraeidae 6, 14, 21, 23, 27, **152**

BERAEODES 21, **154**

Bicolor Sedge 162

bicolor, Triaenodes 22, **162**, 179

bilineatus, Athripsodes 22, **165**, 166

*bilineatus, Leptocerus* 165

bimaculata, Neureclipsis 17, **62**

binotatus, Limnephilus 20, **130**, 131

bipunctata, Phryganea 18, **102**

bipunctatus, Limnephilus 20, **131**

Black Sedges 105

Black Silverhorns 164, 168, 174, 175

boltoni, Glossosoma 15, **31**

*boltoni , Glossosoma* 16, **32**

borealis, Limnephilus 20, **127**, 128

Brachycentridae 5, 14, 19, 27, **102**

BRACHYCENTRUS 3, 19, **103**

brevipennis, Anabolia 19, 117, **120**, 140

*brevipennis, Phacopteryx* 120

brevis, Plectrocnemia 17, **71**

Brown Sedge 119

Brown Silverhorns 164, 168

bulgaromanorum, Hydropsyche 18, **88**

*CABORIUS* 113

centralis, Limnephilus 20, **135**

CERACLEA 22, 159, 164, **168**

*cesareus, Philopotamus* 56

Chaetopterygini 19, **140**

CHAETOPTERYX 19, **140**

CHEUMATOPSYCHE 18, 82, **83**

CHIMARRA 17, **54**

*CHIMARRHA* 55

*chrysopterus, Philopotamus* 56

ciliaris, Notidobia 21, **151**

cinereus, Athripsodes 22, **166**

*cinereus, Leptocerus* 166

cingulatus, Potamophylax 21, **144**, 145

Cinnamon Sedges 128, 129

clathrata, Hagenella 10, 18, **98**

*clathrata, Neuronia* 98

clavata, Ithytrichia 16, **44**

*coenosus, Asynarchus* 139

coenosus, Limnephilus 20, **139**

*COLPOTAULIUS* 125

*comatus, Agapetus* 15, 34

commutatus, Athripsodes 22, 164, **166**

*commutatus, Leptocerus* 166

*concentricus, Stenophylax* 21, 143

conformis, Glossosoma 16, **32**

conspersa, Plectrocnemia 17, **71**

*conspersus, Triaenodes* 163

conspersus, Ylodes 22, **163**

contubernalis, Hydropsyche 18, **88**

cornuta, Hydroptila 16, **41**

costalis, Orthotrichia 16, **48**, 49

*costalis, Oxyethira* 16, **50**

crassicornis, Agrypnetes 18, 90, **93**

CRUNOECIA 19, **107**

CYRNUS 17, 61, **65**


dalecarlica, Ecclisopteryx 19, **115**

Dark Peter 95

Dark Spotted Sedges 61

*DASYSTEGIA* 94, 95

decipiens, Limnephilus 20, **131**

delicatulus, Agapetus 15, **35**

Dicosmoecinae 19, **113**

digitatus, Halesus 20, **147**

DIPLECTRONA 8, 18, **82**

dissimilis, Ceraclea 22, **171**

*dissimilis, Leptocerus* 171

distinctella, Oxyethira 16, **53**

dives, Tinodes, 18, **78**

dorsalis, Rhyacophila 15, **27**

Drusinae 19, 112, **114**

DRUSUS 19, 112, **114**

dubia, Ironoquia 19, **113**

dubitans, Synagapetus 15, 16, **35**

*dubius, Caborius* 113

dubius, Holocentropus 17, **63**


ECCLISOPTERYX 19, 112, 114, **115**

Ecnomidae 14, 17, 25, **60**

ECNOMUS 17, **60**

elegans, Limnephilus 20, **132**

ENOICYLA 20, 118, **148**

ERNODES 21, **155**

EROTESIS 22, 159, **160**

exocellata, Hydropsyche 5, 18, **89**

extricatus, Limnephilus 20, **135**, 136


*fagesii, Leiochiton* 49

fagesii, Tricholeiochiton 17, **49**

falcata, Oxyethira 16, **52**

fasciata, Rhyacophila 15, **28**

felix, Diplectrona 18, **82**

*femoralis, Hydroptila* 16, **37**

filicornis, Adicella 21, **160**

*fimbriata, Apatania* 110

*fimbriata, Apatidea* 110

*fimbriatum, Lepidostoma* 19, 108

*flavicorne, Sericostoma* 150

flavicornis, Limnephilus 20, **127**

flavicornis, Oxyethira 16, **50**

flavidus, Cyrnus 17, **66**

flavomaculatus, Polycentropus 17, **68**, 70

forcipata, Hydroptila 16, **38**

*fragilis, Metalype* 76

fragilis, Psychomyia 17, **76**

frici, Oxyethira 16, **50**, 52

fulva, Ceraclea 22, **169**

fulvipes, Hydropsyche 18, 84, **86**

*fulvus, Leptocerus* 169

furva, Oecetis 22, **179**

fuscicornis, Limnephilus 20, **134**

fuscinervis, Limnephilus 15, 20, **134**
fuscipes, Agapetus 15, **34**, 35

geniculata, Plectrocnemia 17, **72**
GLOSSOSOMA 6, 15, **30**
Glossosomatidae 14, 15, 26, **30**
Glossosomatoidea 15
GLYPHOTAELIUS 19, 116, **121**, 123
GOERA 19, **104**
Goeridae 5, 14, 19, 25, **104**
GRAMMOTAULIUS 20, 116, **122**
grandis, Phryganea 18, **101**, 102
Grannom 103
Great Red Sedges 100
Greentail 103
Grey Flags 84
Grey Sedge 156
griseus, Limnephilus 20, 131, **132**
Grouse Wing 176
*guttata, Hydropsyche* 18, **88**
*guttatipennis, Halesus* 148
*guttatipennis, Melampophylax* 21, 148
*guttulata, Ecclisopteryx* 15, 115

HAGENELLA 18, 91, **98**
HALESUS 20, 119, **147**
hirsutus, Limnephilus 20, 135, **136**, 137
hirtum, Lepidostoma 19, **108**
HOLOCENTROPUS 17, 62, **63**
HYDATOPHYLAX 20, 118, **149**
HYDROPSYCHE 2, 4, 18, 81, 83, **84**
Hydropsychidae 14, 18, 26, **81**
Hydropsychoidea 17
HYDROPTILA 16, 36, **37**
Hydroptilidae 14, 16, 23, **36**
Hydroptiloidea 16

ignavus, Limnephilus 20, **133**
impunctatus, Mesophylax 21, **141**
incisus, Limnephilus 20, 117, **125**
*incisus, Colpotaulius* 125
infumatus, Hydatophylax 20, **149**
*infumatus, Stenophylax* 149
insolutus, Cyrnus 17, **67**

instabilis, Hydropsyche 18, **84**, 85
*insularis, Philopotamus* 56
Integripalpia 14, 18
*interjectus, Athripsodes* **164**, 165
*intermedia, Mystrophora* 32
intermedium, Glossosoma 16, **32**
*interrupta, Setodes* 173
interruptus, Leptocerus 22, **173**
IRONOQUIA 19, 112, **113**
irrorata, Crunoecia 19, **107**
irroratus, Polycentropus 17, **69**
ITHYTRICHIA 16, 36, **44**

kingi, Polycentropus 17, **70**

lacustris, Oecetis 22, **179**
lamellaris, Ithytrichia 16, **44**
Large Cinnamon Sedges 143, 144
Large Red Sedges 100
*LASIOCEPHALA* 19, 108, **109**
lateralis, Micropterna 21, **142**
*lateralis, Stenophylax* 142
latipennis, Potamophylax 21, **145**
*latipennis, Potamophylax* 21, 144
*latipennis, Stenophylax* 144
*LEIOCHITON* 49
lepida, Cheumatopsyche 18, **83**
LEPIDOSTOMA 19, 107, **108**
Lepidostomatidae 5, 6, 14, 19, 26, **107**
Leptoceridae 3, 8, 14, 21, 23, 27, **158**
Leptoceroidea 21
LEPTOCERUS 22, 159, **172**
Limnephilidae 2, 5, 7, 10, 14, 19, 24, 110, **112**
Limnephilinae 19, 112, 113, **116**
Limnephilini 19, **119**
Limnephiloidea 19
LIMNEPHILUS 20, 117, **124**
Longhorn Sedges 178
longicornis, Mystacides 22, **176**
*longispina, Hydroptila* 16, **37**
lotensis, Hydroptila 16, 40, **41**
lunatus, Limnephilus 20, 127, **128**, 129
luridus, Limnephilus 20, **133**
*lusitanica, Setodes* 173

lusitanicus, Leptocerus 22, **173**
LYPE 17, 72, **73**

maclachlani, Tinodes 18, **80**
*maclachlani , Hydroptila* 16, **38**
maculicornis, Tinodes 15, 18, **79**
Marbled Sedges 84
marginata, Chimarra 17, **55**
marmoratus, Limnephilus 20, 126, 128, **129**
martini, Hydroptila 16, **43**
maurus, Beraea 21, **152**, 153
mediana, Wormaldia 17, **58**, 59
Medium Sedge 104
MELAMPOPHYLAX 21, 119, **148**
MESOPHYLAX 21, 118, **141**
*METALYPE* 17, 75, **76**
Micro Black Sedges 73
Micro Caddis 36
MICROPTERNA 21, 118, **142**
*minor, Nannophryganea* 99
minor, Trichostegia 18, **99**
minutus, Beraeodes 21, **154**
*mirabilis, Oxytrichia* 52
mirabilis, Oxyethira 16, **52**
MOLANNA 21, **157**
Molannidae 14, 21, 26, **157**
montanus, Philopotamus 17, **56**
Mottled Sedge 121
mucoreus, Melampophylax 21, **148**
muliebris, Apatania 19, **111**
*muliebris, Apatidea* 111
*multiguttatus, Polycentropus* 17, 68, 69
multipunctata, Agraylea 16, **45**, 46
munda, Rhyacophila 15, **29**
Murraghs 100
MYSTACIDES 8, 22, 159, **174**
*MYSTROPHORA* 32

*NANNOPHRYGANEA* 99
NEMOTAULIUS 20, 116, **123**
nervosa, Anabolia 19, **119**
NEURECLIPSIS 17, 61, **62**
*NEURONIA* 98, 100
*nielseni, Apatania* 19, 111

nigra, Mystacides 22, **175**
nigriceps, Limnephilus 20, **138**
nigricornis, Silo 19, **105**, 106
nigronervosa, Ceraclea 22, **170**
*nigronervosus, Leptocerus* 170
nigropunctatus, Grammotaulius 20, **122**
nitidus, Grammotaulius 20, **122**
notata, Oecetis 22, **180**
NOTIDOBIA 21, 150, **151**

obliterata, Rhyacophila 15, **29**
obsoleta, Agrypnia 18, **95**
*obsoleta, Dasystegia* 95
*obsoleta, Phryganea* 95
occipitalis, Wormaldia 17, **57**, 58, 59
occulta, Hydroptila 16, **42**, 43
ochracea, Oecetis 22, **178**
ochripes, Agapetus 15, **34**
Odontoceridae 14, 21, 26, **156**
ODONTOCERUM 21, **156**
OECETIS 22, 159, **178**
OLIGOTRICHA 18, 91, **100**
*ornatula, Hydropsyche* 18, **88**
ORTHOTRICHIA 16, 36, **47**
OXYETHIRA 16, 36, **50**
*OXYTRICHIA* 52

pagetana, Agrypnia 18, **97**
pallicornis, Allotrichia 16, **46**
*pallidula, Agraylea* 16, **46**
pallidulus, Tinodes 18, **81**
pallipes, Silo 19, 105, **106**
*palpata, Molanna* 21, 157
pati, Limnephilus 20, 136, **137**
pellucidula, Hydropsyche 18, **87**
pellucidus, Glyphotaelius 3, 19, **121**
permistus, Stenophylax 21, **143**, 144
personatum, Sericostoma 21, **150**
*PHACOPTERYX* 19, 119, 120
phaeopa, Lype 17, **73**, 74
Philopotamidae 14, 17, 26, **54**
Philopotamoidea 17
PHILOPOTAMUS 17, 54, **55**
PHRYGANEA 18, 91, **100**

Phryganeidae 5, 7, 14, 18, 24, **90**

Phryganeoidea 18

picicornis, Holocentropus 17, **64**

picta, Agrypnia 18, **96**

pilosa, Goera 19, **104**

PLECTROCNEMIA 17, 62, **70**

politus, Limnephilus 20, **130**

Polycentropodidae 8, 14, 17, 25, **61**

POLYCENTROPUS 17, 61, **68**

POTAMOPHYLAX 21, 143, **144**

PSYCHOMYIA 17, 72, **75**

Psychomyiidae 14, 17, 26, **72**

pulchricornis, Hydroptila 16, **37**

pullata, Beraea 21, **153**

punctatolineatus, Nemotaulius 20, **123**

punctatus, Setodes 22, **177**

pusilla, Enoicyla 20, 23, **148**

pusilla, Psychomyia 17, **75**

*pusillus, Tinodes* 18, 80

radiatus, Halesus 20, **147**

reducta, Lype 17, 73, **74**

reducta, Adicella 21, **161**

*reuteri, Triaenodes* 163, 164

reuteri, Ylodes 22, **163**, 164

RHADICOLEPTUS 20, 118, **124**

rhombicus, Limnephilus 20, **126**, 129

RHYACOPHILA 2, 15, **27**

Rhyacophilidae 14, 15, 24, **27**

Rhyacophiloidea 15

rostocki, Tinodes 18, **78**

rotundipennis, Potamophylax 21, **146**

*rotundipennis, Stenophylax* 146

*ruficrus, Neuronia* 100

*ruficrus, Oligotricha* 18, 100

sagittifera, Oxyethira 16, **53**

Sand Fly 27

saxonica, Hydropsyche 18, **86**

Scarce Brown Sedge 113

Scarce Grey Flag 88

schneideri, Sericostoma 150

*scoticus, Philopotamus* 56

senilis, Ceraclea 22, **171**

*senilis, Leptocerus* 171

*septentrionis, Rhyacophila* 15, **28**

sequax, Micropterna 21, **142**

*sequax, Stenophylax* 142

SERICOSTOMA 21, **150**

Sericostomatidae 5, 14, 21, 27, **150**

Sericostomatoidea 21

SETODES 22, 159, **176**

sexmaculata, Agraylea 16, **46**

SILO 19, 104, **105**

siltalai, Hydropsyche 18, **85**

Silver Sedge 156

simplex, Oxyethira 16, **51**

simulans, Hydroptila 16, **40**, 41

*simulans, Triaenodes* 164

simulans, Ylodes 22, **164**

*simulans, Ylodes* 22, 163

Small Grey Sedge 32

Small Red Sedge 77

Small Silver Sedge 108

Small Yellow Sedge 75

sparsa, Hydroptila 16, **40**, 41

sparsus, Limnephilus 20, **134**

Speckled Peter 94

Spicipalpia 14, 15

stagnalis, Holocentropus 17, **64**

*stellatus, Potamophylax* 21, 145

*stellatus, Stenophylax* 145

Stenophylacini 20, **141**

STENOPHYLAX 21, 118, 142, **143**

stigma, Limnephilus 20, **129**

striata, Oligotricha 18, **100**

*striata, Phryganea* 18, 102

subcentralis, Limnephilus 20, **128**

subnigra, Wormaldia 17, **59**

subnubilus, Brachycentrus 19, **103**

sylvestris, Hydroptila 16, **39**

SYNAGAPETUS 16, 30, **35**

tauricus, Limnephilus 20, 136, **137**

tenellus, Ecnomus 17, **60**

testacea, Oecetis 22, **180**

*tetensii, Orthotrichia* 16, **48**

tigurina, Hydroptila 16, **39**

tineiformis, Leptocerus 22, **172**, 173

*tineiformis, Setodes* 172

tineoides, Hydroptila 16, **37**

TINODES 18, 72, 77

Tiny Grey Sedges 33

tragetti, Orthotrichia 15, 16, **49**

TRIAENODES 22, 159, **162**, 163

TRICHOLEIOCHITON 17, 36, **49**

TRICHOSTEGIA 18, 91, **99**

trimaculatus, Cyrnus 17, **66**, 67

tristella, Oxyethira 16, **51**


unicolor, Tinodes 18, **79**


valesiaca, Hydroptila 16, **43**

varia, Agrypnia 18, **94**, 95

*varia, Dasystegia* 94

*varia, Phryganea* 94

vectis, Hydroptila 16, **38**

*vernale, Glossosoma* 15, **31**

vibex, Stenophylax 21, **144**

villosa, Chaetopteryx 19, 117, **140**

vittatus, Limnephilus 20, **139**


waeneri, Tinodes 18, 77

wallengreni, Apatania 19, **111**

Welshman's Button 150

Window Winged Sedge 98

WORMALDIA 17, 54, **57**


*xanthodes, Limnephilus* 20, 130


Yellow Spotted Sedge 56

YLODES 22, 159, **163**


zetlandicus, Mesophylax 21, 141